Ernest Ingersoll

To The Shenandoah And Beyond

Ernest Ingersoll

To The Shenandoah And Beyond

ISBN/EAN: 9783741118210

Manufactured in Europe, USA, Canada, Australia, Japa

Cover: Foto ©Andreas Hilbeck / pixelio.de

Manufactured and distributed by brebook publishing software (www.brebook.com)

Ernest Ingersoll

To The Shenandoah And Beyond

TO THE

SHENANDOAH AND BEYOND:

THE CHRONICLE OF A LEISURELY JOURNEY

Through the Uplands of Virginia and Tennessee,

SKETCHING THEIR SCENERY, NOTING THEIR LEGENDS,
PORTRAYING SOCIAL AND MATERIAL PROGRESS,
AND EXPLAINING ROUTES OF TRAVEL.

BY
ERNEST INGERSOLL.

With Illustrations by
FRANK H. TAYLOR.

NEW YORK:
LEVE & ALDEN PRINTING COMPANY, 107 LIBERTY STREET.
1885.

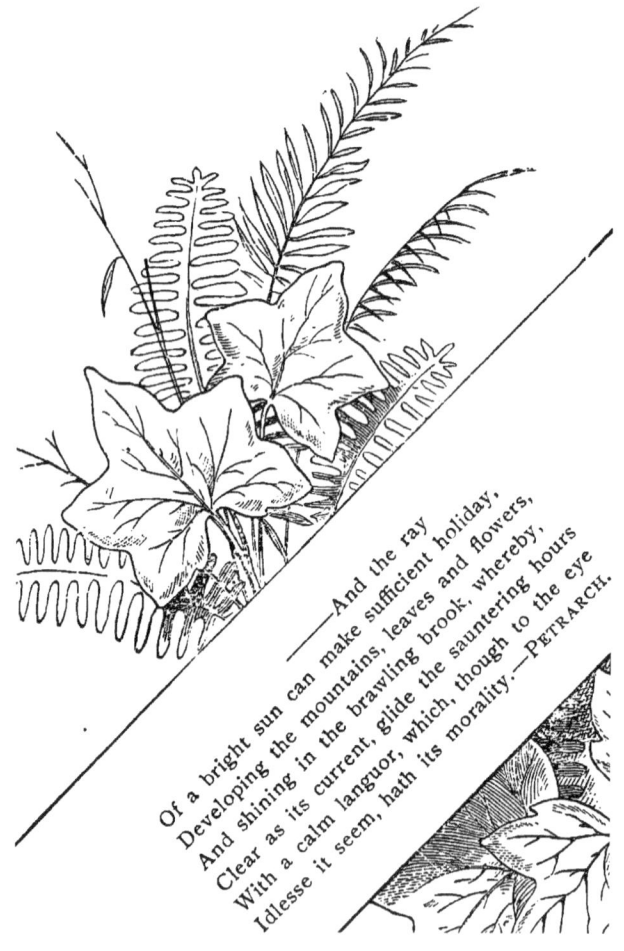

And the ray
Of a bright sun can make sufficient holiday,
Developing the mountains, leaves and flowers,
And shining in the brawling brook, whereby,
Clear as its current, glide the sauntering hours
With a calm languor, which, though to the eye
Idlesse it seem, hath its morality.—PETRARCH.

TABLE OF CONTENTS.

	PAGE
I.—THE CUMBERLAND VALLEY,	5
II.—IN AND ABOUT HAGERSTOWN,	10
III.—PEN-MAR AND BLUE MOUNTAIN,	13
IV.—ON THE WESTERN MARYLAND RAILWAY,	16
V.—THE ANTIETAM AND THE POTOMAC,	19
VI.—THE LOWER VALLEY OF THE SHENANDOAH,	24
VII.—LURAY AND ITS CAVERNS,	31
VIII.—UP THE SOUTH FORK,	42
IX.—CRAB-TREE FALLS AND THE NATURAL BRIDGE,	47
X.—THE NEW CITY OF ROANOKE,	56
XI.—NORFOLK AND PETERSBURG,	69
XII.—IN THE VALLEY OF THE JAMES,	82
XIII.—WESTWARD BOUND,	88
XIV.—NEW RIVER CAÑON AND MOUNTAIN LAKE,	92
XV.—THROUGH SOUTH-WEST VIRGINIA,	98
XVI.—ROAN MOUNTAIN AND THE CAÑONS OF DOE RIVER,	102
XVII.—THROUGH EAST TENNESSEE,	106
XVIII.—A CHAPTER EXPLANATORY,	113

LIST OF ILLUSTRATIONS.

	PAGE
Baily gets the Facts,	6
Hagerstown,	11
Hagerstown Station,	12
Evening on the Upper Potomac,	22
A Dot of a Cabin,	25
Virginia Uplands,	27
Between Front Royal and Luray,	29
Hall of the Giants,	32
Banks of the Rhine, "Luray Caverns,"	35
A Mountain Cascade,	37
Luray Inn,	38
Station and Restaurant at Luray,	39
An "Interior" in the Inn at Luray,	41
The Blue Bridge, near Waynesboro,	45
The Natural Bridge,	50
The Saltpetre Cave on Cedar Creek,	52
The Arbor Vitæ Trees and Giants' Stairway,	55
James River Gorge,	57
Near Buchanan,	58
Crozier Iron Works,	59
A Mountain Rift near Roanoke,	60
Offices of the Consolidated Railways at Roanoke,	62
Big Spring, near Roanoke,	63
Hotel Roanoke,	64
Tinker and Mill Mountains, Roanoke,	65
Hotel Roanoke,	66
Lobby of the Hotel Roanoke,	67
The Market Square at Norfolk,	69
New Cotton Compressor at Norfolk,	70
Old Church at Norfolk,	71
Terminal Wharves at Lambert's Point,	72
In Fortress Monroe,	73
Old Point Comfort, from Soldiers' Home,	74
Views from Dock Pavilion, Hotel Warwick,	75
Bowling Hall, Hotel Warwick,	76
Scene on Virginia Beach,	77
Hampton Roads,	78
New Station of the Norfolk & Western Railway at Norfolk,	79
A Suffolk Farm House,	80
Conoeing on the Dismal Swamp,	80
Footprints of War,	81
Tobacco Wagons at Lynchburg,	84
Negro Wagoners,	84
An Ebony Gabriel,	84
The Peaks of Otter,	85
The Roanoke,	89
New River Scenery,	94
The Maple Shade Inn,	98
On Doe River,	103
Along the Upper French Broad,	107

INDEX.

[See also Appendix.]

	PAGE
Abingdon,	101
Air Line,	9, 61
Alabama cities,	114
Antietam, battle of the,	23, 24
Appalachian Valley, the Great,	6
Appomatox station,	83
Asheville,	108
Atlanta,	115
Baltimore and Ohio railway,	24
battle of Antietam, —.	23, 24
—— Buchanan,	57
—— Cross Keys, —.	43
—— Front Royal, .⁻	30
—— Knoxville,	110
—— Pack-horse ford,	23
—— Port Republic, ⁓	42
—— South-west Virginia,	91, 93, 98, 101
—— Waynesboro,	46
Berryville,	26
Blue Mountain house,	13, 14
" Blue Mountain route,"	115
Blue Ridge,	14, 25, 45
Blue Ridge Springs,	87
Bluestone,	91
Bristol,	101
Buchanan,	56
Carlisle,	8
Castle Cresap,	10
Cattle in South-west Virginia,	64
Caverns of Luray,	31-42
Chambersburg,	9
Chapel, the "old,"	26
Charleston,	24
Chattanooga,	112
Chesapeake bay steamers,	63
Chesapeake & Ohio railroad,	46
Christiansburg,	91
Cleveland,	112
Cloyd's mountain,	92
Corn-Exchange regiment,	24
Cotton traffic at Norfolk,	71
Crab-tree Falls,	47, 48
Cranberry iron mines,	102
Cripple Creek,	98
Cross Keys, ⁓.	43
Crozier Steel and Iron Company,	64
Cumberland valley,	5, 7, 15

	PAGE
Dismal Swamp,	78
Doe river,	102
Elkton,	42, 44
East Tennessee,	103, 110
East Tenn. & N. Carolina R'y,	103
East Tennessee, Virginia & Georgia R'y,	106, 112
Fairfax's estate,	27
Farmville,	82
Flat-top Coal mines,	91
Florida,	115
" Florida Short line,"	115, 116
Forks of the Shenandoah,	28
Fortress Monroe,	73
French Broad river,	108
Front Royal,	29-31
Gettysburg,	8
Greenville,	107, 109
Greenway court,	27
Grimes station,	20
Hagerstown,	10-12, 20
Hampton Roads,	76
Harewood House,	25
Harrisburg,	5, 6
Hawksbill valley,	42
High Bridge,	82
Hotel Roanoke,	67
Hotel Warwick,	75, 76
Hunter's raiders,	58, 87
Hygeia hotel,	74
Indians,	10, 20, 23, 29
Iron-ore,	30, 44, 64, 98, 101
Jackson, Gen. Stonewall,	21, 25, 42, 49
James river,	49, 56, 59, 83
Jellico,	109
Knoxville,	109
Lambert's Point,	72
Lexington,	49
Liberty,	86
Loch Laird,	49
Luray and its caverns,	33-40
—— history of,	41
Luray Inn,	40
Lynchburg,	83, 84

INDEX.

	PAGE
Maple Shade Inn,	99
Marble of Tennessee,	107
Martins,	99
Massanutten mountain,	28
Max Meadows,	99
Memphis,	114
Milnes station and forges,	44
Minerals of Blue Ridge,	30, 42, 44
—— Southwest Virginia,	101
Mobile,	114
Mountain-lake ("salt-pond"),	96
Natural Bridge,	49-56
New Orleans,	114
Newport News,	76
New river,	91-97
Norfolk,	69
Norfolk and Western railway,	61, 69
Oakvale,	95
Ocean View,	76
Old Point Comfort,	74
Pack-horse ford,	23
Page valley,	31
Pamplin's depot,	83
Peaks of Otter,	84, 87
Peanut culture in Virginia,	72
Pearisburg,	92
Pen-Mar,	13
Petersburg,	80
Pike, the national,	11
Pocahontas,	91, 95
Port Republic,	42
Potomac river,	21, 22
Quirauk,	14
Railway, Cumberland Valley,	5, 19
Railway routes southward,	112, 114
—— westward,	114
Ramsay's steamboat,	21
Richmond and Alleghany railway,	49, 56
Ringgold's Manor,	19
Riverton,	28
Roan mountain,	105
Roanoke, city of,	59-64
—— machine works,	63
—— valley,	59

	PAGE
Salem,	88
Saltville,	100
St. James,	19
Saratoga house,	27
Sharpsburg,	20
Shenandoah Iron Works,	44
—— Junction,	24
—— river,	7, 29, 43, 46
—— Valley Railroad,	9, 19, 26, 61
Shepherdstown,	21-24
Sheridan-Early campaign,	25, 26, 30, 43, 46
Southern Fast Mail,	16
South Fork of Shenandoah,	42-43
Spottswood and the "Golden Horseshoe,"	43
Springs, Alleghany,	90
—— Bedford Alum,	87
—— Blue Ridge,	87
—— Eggleston's,	93
—— Farmville Lithia,	83
—— New River White Sulphur,	91
—— Red Sulphur,	88
—— South-west Virginia,	98
—— Warm, in N. Car.,	109
—— White Sulphur,	91
—— Yellow Sulphur,	91.
Swamp, great dismal,	78
Tennessee river,	110
Trout-fishing,	96
Valley of Virginia,	25
Virginia Beach,	77
Virginia Central railroad,	46
Virginia Midland crossing,	28
Virginia, southern,	80
Washington's explorations,	27
Washington, Ohio and Western railway,	26
Waynesboro,	46
—— Junction,	45
Western Maryland railway,	13-19
Western North Carolina,	105
White Post,	27
White Sulphur Springs,	45
Williams' Grove,	8
Wytheville,	99

TO THE

SHENANDOAH AND BEYOND.

I.

THE CUMBERLAND VALLEY.

A Railway Centre.—Astonishing Business of Harrisburg Station.—Æsthetic and Antiquarian Attractions at Harrisburg.—The "Appalachian Valley."—Williams' Grove.—Carlisle and its Schools.—Snugness and Prosperity of the Towns.—Fertility and High Cultivation of the Soil.—Chambersburg and its History.—Gettysburg Battlefield.—Maryland.

ALTHOUGH a fairly well-informed young person, Prue sometimes asks very singular questions. An example occurred this very morning.

We were waiting beside a little heap of valises and rolls of wrappings in the railway station at Harrisburg, Pennsylvania, when Prue, turning questioning eyes up to mine —very pretty eyes they are, too, I think—remarked:

"Theodore, what is meant by the phrase 'a railroad centre?'"

"Eh?—why—*this* is one: A point where various lines of railways converge and concentrate their forces."

"How is Harrisburg an example?"

"Well, it is a sort of half-way house on the great Pennsylvania Road east and west, and the place where its principal branch, the Philadelphia and Erie, diverges. Here crosses, also, the north and south trunk route connecting Baltimore and Washington with all the cities of western New York state and the Canadas. This is the end of the Philadelphia and Reading, and, finally, it is the gateway of the South, since here begins the Cumberland Valley Railroad, by which we are to begin our travels through the Southern mountains. You could hardly find a spot in the eastern and middle states where the straightest practicable course to the chief southern cities would not lead through Harrisburg; or, for persons going northward, one where they could find more conveniences for reaching their diverse destinations. Now do you see why it is called a railway centre?"

"Indeed I do. What a busy place this station must be, for all the trains seem to come into this one building."

"Yes, I daresay—Oh, Baily, suppose you go and ask the stationmaster how many trains go and come here each day."

"Er—I was just going to——"

"No matter about it, now, whatever it was, but just go along, like a good fellow."

Prue was the last speaker, and was irresistible. Baily went. I knew he would. We had asked him, an idler, to come with us, because he had so obliging a disposition. I felt rather mean about it, for I knew how he would be imposed upon, but Prue declared he liked it, and I let her take the responsibility.

BAILY GETS THE FACTS.

Presently Baily came back, flourishing a large Russia-bound, gilt-edged memorandum-book, opened at the first page.

"Couldn't remember it all without notes," he ejaculated, and began to read statistics.

"Stop! stop!" I cried after a short siege of this, while Prue stood aghast at the figures she had summoned to arise before her—statistical ghosts that "could not be laid"—because, you know, figures will not *lie.* "Stop! Isn't there any general result? What is the sum?"

"Oh, yes," Baily answered with unfailing cheerfulness, "about seventy-eight passenger trains go through the station every twenty-four hours, and each has seven or eight cars; while the innumerable freight trains bring nearly 3,000 cars more a day into this yard. When I add that a hundred men are employed in station duties I am done," and the Russia book was closed gently.

It was just in time. The Cumberland Valley train was ready for us, and a few moments later our jolly tour down the long Appalachian valleys, which stretch from here to Georgia, had begun.

Harrisburg was worth a longer stop than our arrangements allowed. It is a beautiful old city, with a great deal to interest the visitor. The central show-place, of course, is the state' capitol, set upon a hill in the midst of a highly-cultivated park. Though of brick, and according to a style now out of vogue, it is a dignified and commodious building. Near it is the fine Corinthian shaft, bearing a noble figure, which commemorates the dead soldiers of the Mexican war. It is surrounded by a fence of real muskets and captured cannon, and is one of the most satisfactory monuments of its character in the United States. Only a square away stands the granite obelisk, unadorned, which speaks Pennsylvania's gratitude to her defenders during the late war. The antiquarian and lover of history will probably find more valued and inspiring relics of Colonial and Revolutionary times stored in the public library at Harrisburg than in any other state house in the country; while many valuable paintings adorn its alcoves.

Harrisburg is wealthy and aristocratic, as its homes testify. We get a glimpse of the exterior of a group of the best of these as the train moves out upon the bridge to cross the Susquehanna. A long street fronts the river, with square after square of noble houses and charming gardens. Their outlook is unobstructed, for between the street and the brink of the river-bluff runs only a narrow park.

The river is more than half a mile broad here, shallow and rocky. Forster's island divides it—an island of market-gardens. Our bridge must be nearly a hundred feet above the water, and from it an exceedingly beautiful view is presented, with the warped and mossy old covered bridge, built for wagon traffic in 1812, as a picturesque feature in the foreground.

Having crossed to the southern bank of the Susquehanna, we are at once among rural scenes. Here begins the broad expanse between the Blue Ridge or South Mountain, and the front rank of the Alleghanian ranges, called the North Mountain. So much of this expanse as lies in Pennsylvania and Maryland is called the Cumberland (County) Valley, and its principal stream is the Cónedegwinit.

"It is a splendid series of such valleys that we are to traverse," I say to my companions. "Always on our left, and southeast of us, the chain of the Blue Ridge; always on the right hand and northward, the bold front of the Alleghanies. Sometimes, as here, these ranges will be thirty miles apart. Sometimes they will come close together. Below this valley lies that of the Shenandoah. Beyond that the basin of the Roanoke, and so on. Sometimes we shall turn aside into the mountains themselves, or stop to rest beneath their shadow."

"What a lovely time we are to have!" exclaimed Prue delightedly, and tucked her hand in sign of gratitude within my arm; that was reward enough for anybody.

Meanwhile we were out among the farms—forests of maize, the tassels and flag-like leaves nodding and snapping under the breeze raised by our swift passing; yellow spaces of stubble, where acres upon acres of grain have stood; blossoming squares of clover; wide meadows of pasture and hay; emerald fields of tobacco.

Though in so high a state of cultivation, there was a fair proportion of woodland, and trees of great size grew in the fence corners and about the houses. One might consider these relics of the primeval forest, but none of them were older than the settlements, for when the whites first came to this region, "the whole extent of country between the South and Blue mountains, from the western bank of the Susquehanna to Carlisle, was without timber."

Everything looked prosperous. Not a hovel was to be seen—not even a poor, untidy little cabin on some bare knoll. Every house seemed to be a homestead which had descended from the fathers, as, in perhaps the majority of cases, was really the fact. Two or three villages flitted by, and a junction, whence, Baily informed us, a branch line led to Williams' Grove.

"What is that!"

"A sequestered and sylvan retreat beside a murmuring stream, or something of that sort," says Bailey, "where the National Grange has held an annual assemblage for so many years that it has now become perennial. It is a sort of vast picnic, camp-meeting, agricultural fair and political mass meeting, all mixed into one, and a hundred thousand persons visit it every year. There's a heap o' fun to be had there among the grangers and their girls!" exclaims the volatile Baily.

A little later we came to another junction. Our glib companion did not stop—he simply changed his subject.

"That's the line to Gettysburg—only an hour's ride. Mighty pretty country, and lots of people run down there to see the famous battle-ground, and to kick up the dust searching for bullets."

Off to the right and ahead of the junction, a lot of low white buildings, in the midst of wide and highly cultivated farm lands, next attracted attention.

"Hello!" was his remark, "There's the government school for Indian children."

"Ah, yes; I read a long article describing it in *Harper's Magazine* for April, 1881. Four or five hundred boys and girls are gathered there from all sorts of tribes and every part of the country where the red man now remains, and are turned out good farmers, craftsmen and housekeepers, to become teachers and exemplars to their own people. This town we are entering, must therefore be Carlisle."

Carlisle it is—one of the oldest settlements in the region. The railway runs right along the middle of the principal street, where the best houses, the public buildings and the finest stores are situated. Thus one gets quite as good an idea of the neat little city as if he rode about it for a week. Rooms in an ordinary building serve as station offices, and passengers alight from the train almost upon the sidewalk in front of the principal hotels—a home-like welcoming arrangement, giving a pleasant impression.

Just beyond the stopping-place, on the right, stand the spacious grounds and buildings of Dickinson College, where from 100 to 150 students gather annually. One of its early professors has become a very famous man—Spencer F. Baird, Secretary of the Smithsonian Institution and United States Commissioner of Fisheries.

Beyond Carlisle the country became more rolling and better wooded. Wide areas of clover appeared, and the roadsides were richly blue for a little way with an aster-like blossom in the greatest profusion, beside which grew the sulphur-yellow heads of toad-flax. Enormous corn fields undulated over the knolls, and nowhere was there to be seen a particle of swampy, or stony or waste land. Every rod paid toll in produce or pasturage, timber or fuel, to the thrifty husbandmen. The very towns, Newville, Shippensburg, Chambersburg (a large, active place, with a good deal of manufacturing), Greencastle, and the smaller ones between, all betray the same feeling. They are compact and well-kempt. The houses stand in blocks and are set flush upon the street. Each shines with neatness and clean paint, and in the little garden behind you can hardly walk for the trees, bushes and vines bearing fruit, or the vegetables, set as thickly as they will grow. Land is very valuable, and must not be wasted.

These villages and the rich farms were an object of great longing to the Confederate captains whose forces lay so short a distance southward, and especially to the cavalry champions of the Shenandoah valley. During the early part of the great civil conflict, whose bloodiest battlefields we are destined to see, the Union line of defense along the Potomac was too strong to make feasible any raiding northward of that river; but at the end of July, 1864, when Early was making so strong a demonstration on the lower Shenandoah, and before the cavalry army of Sheridan had been formed to check and ultimately to defeat him, more than one rebel raid was executed into this region.

The most extensive and dreadful of all these descents was McCausland's celebrated cavalry dash, wherein he passed behind the Union lines at Hagerstown, and suddenly appeared in Chambersburg, with a demand for $100,000 in gold instantly as a ransom, failing which he would burn the town. It was impossible to pay the ransom, and the town was defenceless. A few moments, therefore, saw the torch applied.

Meanwhile detachments had gone northward even as far as Carlisle, and they will show you there two or three marks of its brief bombardment; but a large Union force going quickly after the raiders, they retreated westward out of the valley, doing such harm as was possible in their flight. In one respect this great fire was a blessing to the burned town, for from its ashes rose a new city far superior in appearance and conveniences.

Near Chambersburg branch railways diverge to Waynesboro, and to Loudon, Richmond and Mercersburg. The main line continues to Martinsburg on the Baltimore and Ohio Railroad, but its practical terminus is at Hagerstown, Maryland. Here terminates, also, the Western Maryland Railway, going to Baltimore; and the Shenandoah Valley road stretching southward as the second link of that great Virginia, Tennessee and Georgia Air Line system, which is better known as " The Shenandoah Route."

II.

IN AND ABOUT HAGERSTOWN.

Traditions of Early Settlements in Western Maryland.—Fort Cresap.—Founding Hagerstown.—Extraordinary Fertility of the Antietam Valley.—The National Pike.—"Public Tuesday."—Picturesque Peculiarities.—War History of the Neighborhood.

THOUGH Hagerstown is in Maryland, and a young lady there was offended when I mentioned what I am about to say, our feeling was, that we had not yet left Pennsylvania. This is due to the fact that the earliest settlers, like the pioneers into the valley of the Susquehanna, were Germans; and they brought with them a certain method and architecture which stamp their settlements, irrespective of state lines, all the way from the Delaware to the Potomac.

It was early in the last century when Europeans first invaded this smiling valley of the Antietam. Tradition says the earliest one of all was that John Howard, who is credited with being the first white man to see the Ohio, and who went thither by this route. Then came and settled here a group of hardy frontiersmen, of whom Thomas Cresap became famous as an Indian fighter in a community always at war with the "red varmints,"—Shawnees, Catawbas or Delawares. The central settlement was at Long Meadows, about three miles from Hagerstown, where a fort of logs and stone was built.

"That was Castle Cresap, was it not?" Asks Prue, whose interest in history never flags.

"The same," I reply. In 1730 there came to the district, from Germany, a Captain Hager, who made a home a mile east of Castle Cresap, which he labeled Hager's Delight. He must have been a man of more than ordinary abilities, I think, since he had previously received a patent for lands now within the city of Philadelphia. Hager's first home was on the Antietam, and was a building of logs having an arched cellar of stone, to which the family would retire whenever they were attacked by Indians.

By 1762 there were people enough in the neighborhood to suggest to Hager the idea of having a town-centre. So he laid off streets across a piece of swampy poor land, and called the plot Elizabethtown, after his wife.

"It seems to have been an unusually good class of immigrants who came in here," Prue remarks (Baily has gone to sleep), "for I remember reading many a stirring incident of Revolutionary story in which these people had a part."

By the beginning of the present century, Elizabeth Hager's Town, or simply Hagerstown, as it was beginning to be called, had become an important centre. The valley was noted for its great crops. There were fifty grist mills in the neighborhood, three iron mines, and half a dozen furnaces and forges, where pig-iron was cast and bars and hollow-ware were made. Markets, churches and substantial houses of brick or stone, many of which remain, had built up the town, whither nearly all the business came, and about which the first macadamized roads in America had been laid down. Large plantations had been organized by slave labor, and civilization was radiating farther and farther into the mountains.

Some outlet was required; better means of getting products out to the seaboard and merchandise in; improved routes of communication with the seaboard. To fill this want the National Pike was begun. Properly speaking, the Pike ran from Cumberland to Wheeling, since only that part of it was built by the Federal government; the part from Cumberland to Baltimore having been constructed under law by certain Maryland banks, which found its tolls a source of great profit. Its glory has departed, but when coaching days were palmy no other post road in the country did an equal business. "The wagons were so numerous," says Howard Pyle, in an article upon it (*Harper's Magazine*, November, 1879), "that the leaders of one team had their noses in the trough at the end of the next wagon ahead, and the coaches drawn by four or six horses dashed along at a speed of which a modern limited express might not feel ashamed."

Many are the good stories which cluster about this portion of that busy highway! Hagerstown has now about 8,500 people, less than five hundred of which are foreign born. If you go there on a Tuesday,

HAGERSTOWN.

however, you will think the whole 35,000 in Washington county have come to town. *Public Tuesday* is an old custom, begun by accident, but now crystallized into a rule of life in that region. It is on that day that courts of record are open; that every business man tries to be at home and every countryman makes his errand to town. That is the time of tax-sales, auctions, hucksterings, cheap shows and everything that seeks a crowd.

Hagerstown is picturesque, well-built and prosperous, with a strong tendency toward manufacturing. From its homes on the hill, where Prospect street asserts its superiority, a wonderfully pleasing picture is presented. The town, with its quaint old houses, many of them of log, more of brick, old-fashioned and embowered in foliage, forms a pretty foreground for the wide space of valley which stretches away to the mountains circling about, in one blue, continuous wall.

Prue found good friends there—where doesn't she? and one evening we sat upon a certain rear porch, eating grapes and talking over the great war-drama enacted in that valley between North and South only a

HAGERSTOWN STATION.

score of years ago. Our host and hostess had seen it all, and the story seemed very real when they could point out each spot and say, "Here, on that hill," or "there, where you see those trees," such and such a scene occurred. The town was seldom, if ever, free from military occupation from beginning to end. Its strategic importance was not great, however, and it was held alternately by northern and southern commanders who had designs elsewhere. Lee's great movement northward toward Gettysburg, and his masterly retreat therefrom, gave to Hagerstown its greatest war incident; but it must not be forgotten that Antietam also was fought within sight and hearing of the village.

III.

PEN-MAR AND BLUE MOUNTAIN.

Baily's noble Self-sacrifice.—How Prue and I ran a Gauntlet.—A Shady Walk to Pen-Mar.—Picnic Arrangements.—Carriages.—Dancing.—The Blue Mountain House.—Mt. Quirauk and High Rock.—Seventy-five Miles at a Glance.

ONE evening we left Hagerstown and went out to spend a day at the Blue Mountain House and Pen-Mar, twenty miles away on the line of the Western Maryland Railroad. The former is a fashionable summer hotel, but Pen-Mar is a great picnic "resort" or pleasure-ground, to which every day of the warm part of the year gather excursions from surrounding towns, but chiefly from Baltimore and its neighborhood. Every few days there arises some occasion when two or three thousand pleasure-seekers gather by special trains and overrun the place. Generally the grounds are nearly deserted by sunset; but now and then an excursion party remains merry-making through the evening.

There were to be some "doings" of this sort that night, and they promised to be worth seeing, while there would be opportunity for Prue to dance to her heart's content, that is, if she were willing to dance with me, since she knew nobody else.

"But there was Baily," you will say to yourself.

Yes, but I told Baily that he couldn't stop, but must go on to Baltimore and come back next morning. Somebody had to. The Western Maryland Railroad must be inspected. Self-sacrifice, always admirable, had to be made, and I nobly said Baily should make it. It is not in me to rob a man of a chance for glory.

So Baily went on, while we stopped at the Blue Mountain House, toiling up the long hill from the station to the breezy heights the house stands upon, running the gauntlet of the curious staring of two or three hundred visitors in holiday dress, who filled the veranda and hallways, and finally reaching our room with much the feeling of a kitten astray at a dog show. But when we had washed our faces and brushed our hair, and Prue had put on that soft, cream-white dress, in which her sweetness is shown to best advantage, then—why then, you know, we were two of the *dogs*—a part of the show—and didn't mind the inquisitive crowd at all.

After tea, we walked half a mile or so through the woods along a smooth and winding path to Pen-Mar. Here the Western Maryland railway, a few years ago, bought an extensive tract of woodland near the top of the South mountain, where their road passes over (or through) it. This is just upon the line between the states of Pennsylvania and Maryland, and the first syllables of each give the combination, which I took to be some strange Cornish or Welsh word, until it was explained.

"I wonder," said that incorrigible Baily, when he heard of it, "why they didn't choose the first and last syllables instead, and so produce *Maria?*"

That's about the size of his taste !

We found a broad and tolerably level area of mountain side from which enough trees and underbrush had been cleared away to give place for roads and buildings, and to allow an outlook. Many small tables and benches were scattered about, where family groups had spread their luncheon. For those who did not bring baskets a large dining hall offered an excellent meal at fifty cents, and at several booths fruit and cooling drinks (but no intoxicants) could be bought. For the little ones swings, whirligigs, a short gravity railway, shooting galleries, bowling alleys and the like were provided at the smallest fees, and two or three photographers flourished by a trade in "tin types."

The finely constructed roads were noisy with carriages, and public hacks carried parties to the top of the mountain and back for ten or fifteen cents a fare. As it was moonlight the hacks were busy even now; but the main evening-attraction was the large covered dancing floor, open on all sides save where the musicians sat.

It was well toward midnight when the whistles called the merrymakers to the last train. Prue said she was glad she was not one of them, but instead might walk pleasantly homeward through the fragrant woods and the glancing moonlight.

The Blue Mountain House is much like many another great summer-season caravansary. It was erected by the Blue Ridge Hotel Company, in 1883, and as it "filled a long-felt want for a first-class resort, within easy reach of Baltimore, Washington and Philadelphia," it was necessary to greatly enlarge it before the opening of the season of 1884. The capacity of the hotel is now 400 guests. The building is splendidly furnished throughout, has large rooms, *en suite* and single, with all modern conveniences, "special regard being paid to the safety, comfort and health of its guests." Sanitary arrangements are carefully planned and constructed; there is soft mountain spring water in abundance, and the premises are lighted by gas. In front of the hotel is an extensive lawn, handsomely laid out and planted with young trees, which some day will grow to be very charming no doubt. Meanwhile there is nothing to obstruct the outlook from the front piazzas across the gnarled tops of the pines at the foot of the lawn, to the wide expanse of the Cumberland valley and the far blue wall of the North mountain.

A pleasant walk of half a mile or so up the mountain brings you to where a rocky crag, called High Rock, rears its head above the forest, and here the proprietors have built an observatory three or four stories in height, where a hundred people at once may sit with the western world at their feet. The crest of the mountain, however, is several hundreds of feet higher and reached by a winding road. On the summit, where the mountain drops steeply away on each side from a narrow ridge, a tower has been built which far overtops the tallest trees.

This is Mount Quirauk (pronounced Quirr-owk), and from it the observer looks both ways and up and down the range. He sees how the Blue Ridge, here, as elsewhere, is really in two lines or double, the more western part being the higher and more continuous. Just here there is a breaking down of the continuity, the depression forming a broad gap, which during the late war was carefully guarded by Federal troops in protection of lower Maryland. The bottom of this gap is cultivated, but elsewhere the mountain is covered with forest.

We stand upon the western summit, here at Quirauk, and looking eastward can distinguish to-day only the misty, prairie-like expanse of central Maryland; but when the air is exceptionally clear, one can detect the shimmer of Chesapeake bay.

Turning our eyes westward and northward, a sharp and varied picture is spread before us under the warm sunlight. On the horizon, so far away as to be vaguely enchanting, lie the folds of Appalachia, rank behind rank. Studying them we can see how the Little North mountain and the Great North mountain overlap ; can pick out certain peaks, and find where " gaps " go through; while southward, just at the end of the picture, are the varied headlands at Harper's Ferry. It is seventy-five miles from one end of that line of mountains to the other !

And what between ? The great smooth plain of the Potomac, known on this side as the Cumberland, and beyond as the Shenandoah valley. It is nowhere perfectly flat, yet nowhere elevated into hills. It is divided into innumerable fields and patches of woodland, whose long-settled boundaries are marked by lines of full-grown trees or by luxuriant hedge-rows, as in a well-ordered park. The varied crops grown—green grass, dull clover, golden stubble, with the warm red-brown of the plowed land and the graceful interference of groves, make an irregular mosaic very pleasant to look upon. In the centre of each cluster of fields stands a white farm-house with its shade trees, its huge barns and surrounding orchards. Here and there, where we can trace the white threads of roads crossing, will be seen a group of such houses and the steeple of a church. At wider intervals, a village with compact masses of brick, and the smoke of factories to distinguish it. Waynesboro, the largest of these, is close at hand, and its great factories for the making of agricultural machinery are in plain view. Hagerstown, twenty miles away, becomes a cluster of spires. The scene is one of agricultural thrift and prosperity, which it would be hard to parallel. Nowhere is there a bit of waste land, nowhere a mean farm or miserable shanty. Everywhere industry and cultivation and general content. No worse shadows lie upon it than those the drifting clouds throw picturesquely down, and the winds as quickly snatch away.

IV.

ON THE WESTERN MARYLAND RAILWAY.

The Fast Mail.—Baily wants to go a-fishing.—Rushing through a Farmer's Paradise.—Pretty Girl at Westminster.—Prue is Shocked.—Another Pretty Girl.—Mountains in Line of Battle.—Trout-brooks and Artists' Foregrounds.—Enthusiasm Justified.

BAILY telegraphed me that he would be back on a train passing Pen-Mar in the early evening, and we resolved to go forward at the same hour. The poor fellow was aghast when he saw us at the station ; but I pointed out to him that there was need to make haste, as it really was of no importance to the world that he should stop at the big hotel or dance with some pretty girl at the picnic grounds. So he lugged his valise back into the car, and we sped away down the hill toward Hagerstown.

"Now," said Prue, kindly wishing to comfort him, "I've no doubt you had a charming trip—tell us all about it."

"Well, I was too vexed at going to take much notice, except that we slid down the long hill with most amazing speed, and then, before I knew it, the mountains were out of sight, and we had rattled through a lot of towns without as much as saying 'By your leave !' and there we were in Baltimore."

"That was the fast mail, my boy. It gathers itself together from Memphis and New Orleans and Atlanta, and all the rest of the far South, at Cleveland, Tennessee, and comes through by the way of the Norfolk and Western, Shenandoah Valley and Western Maryland railways, to Baltimore and Washington in the quickest time ever made on southern railroads."

"I dare say that was the train I returned on."

"No, this is the 'Memphis Express,' but it also is a fast train."

"Fast? Why we shot out of those tunnels from the Union Station, in Baltimore, on Charles street, where all the Northern Central and Pennsylvania trains come and go to New York and Washington, as though we had been fired from a gun ! Then in a minute or two we were scooting along the banks of a little stream which rippled gayly across a pretty meadow, dodging here and there through thickets, in and out of little pools, under foot bridges and over stone barriers, and I wanted to get out and go a-fishing."

"Why didn't you ?"

"Circumstances opposed. First, I hadn't a rod ; second, hadn't a hook ; third, hadn't any bait ; fourth, no fish there—too near town ; fifth, we were running so dam—"

"Oh !" cried Prue in a shocked tone.

"—agingly fast," continued Baily entirely unmoved by her interjection, "that before I could fairly think about it, we had leaped the brook and were racing past orchards red with apples, and fields gray with the stubble of wheat and noisy with the sound of threshers, and every few miles there would be a banging of switches and a rushing by a small station, all the houses of which seemed to rock and dance before our eyes as though a first-class earthquake had put its shoulder under 'em. Then came a big bang and we stopped at Westminster. I tell you it's a pretty country round there! I don't wonder Johnny Reb thought he was in clover when he raided up through Maryland. But we could spare only a minute for Westminster—never saw anybody in such a hurry as was that conductor! There was an awfully pretty girl at the next station, and I spoke to her, and was going to get acquainted, and in a minute more would have kissed her good-by I'm sure—"

"Baily!" says my wife severely.

"Oh, I didn't, Mrs. Prue—really I didn't you know! But before I'd got my lips unpuckered we were where I wanted to go fishing again —and I'd a caught something there, I'm sure. Sometimes this stream would go dashing down little cataracts, and then it would slide along weedy shallows under the willow and sycamore trees, where cattle were standing knee deep in the cool clean current. Paths led up to pleasant farmhouses, and here and there the dammed river—"

"*Mr. Baily,*" Prue breaks in, "is it necessary that you use profane language?"

"Eh? Why I was only saying that dams here and there had raised the water into pretty lakes where an old mill-wheel would be lazily turning and boats were floating, and the summer sky was mirrored blue and still. I came near jumping off. Just think of it! Fishing, swimming and boating, all close by, and I not in it!

"Pretty soon we came to Frederick Junction, where anybody that wanted to could change cars for Frederick, and a lot of other places north and south. The other road ran beneath ours and a young lady got off here, and I was going to help her down the stairs; but just as I was making the arrangement nicely, our train started and I nearly got left."

"That would have been rough," I remarked. "But likely you would have 'got left' in any case."

"N-no," Baily responded slowly, "I'm almost sorry I didn't stick by her. However, there wasn't much time to cry over it. I could see the mountains plainly ahead now—a long wall of 'em and they marched right toward us in line of battle and kept rising higher and higher, while green foot-hills and pretty glens, holding the gay little river, and magnificent farms, flitted past in a swift panorama as though they were all fleeing pell-mell before the advance of the great Blue Ridge. 'T was n't really so, you know," Baily explained, "but it looked so because we were going ahead like the dev—"

Prue sprang to her feet and looked daggers at the excited narrator. I held up my hand and tried to stop his lips, but the effort was too late. Out it came in Baily's full power of voice: "—devotees of the racecourse; and in less than no time we were slap into the mountains, twisting around curves and corners so fast, I thought the engineer was playing 'crack the whip' with us. Then I *did* want to go a-fishing ! Here we were a-calawhooping along a sort of shelf or balcony at the side of a deep ravine, and down at the bottom flowed the most beautiful torrent, dashing and splashing and having a right good time under the trees and among the boulders. I could see pools where I *knew* the trout must be lying, and nobody there to disturb 'em, and I tell you I *did* want to go a-fishing!"

"I don't doubt it! Was the place a pretty one—good for an artist as well as an angler?"

"Pretty? Why, man, old Sonntag himself couldn't ask for a better place to sit and paint! The mountain is cleft by a great irregular gap down which comes this bounding stream. It is wooded everywhere with the most varied and abundant foliage. You look away down into its narrow cliff-walled gorge, and away up to dim heights on the other side, and every little while you can see out beyond, across the Maryland lowlands, where the sunlight is filling the whole world with color and light and cheerfulness. When you have been up that Western Maryland Railroad once, you will resolve that it sha'nt be the last time."

Well, Baily was enthusiastic and talked fast, but he didn't overpraise the comfort and delight of travel by that fine railroad which brings to our southern trunk-line the fast mail from the north, and carries the greater part of the passengers from Baltimore and Washington to the mountain summer resorts, and to the south and southwest.

Its through trains not only come into the depot of the Shenandoah Valley at Hagerstown, instead of into their own, but its coaches run "through" between Roanoke and Baltimore.

V.

THE ANTIETAM AND THE POTOMAC.

Prue Criticises the Author.—Dutch Barns—Ringgold's Manor.—Indian War-paths.—Battlefield of the Antietam.—Lee's Head-quarters.—The Potomac Surprises us.—Shepherdstown.—Recollections of the Bucktails.—Ramsay's Steamboat.—Pack-horse Ford and the Slaughter of the Corn Exchange Regiment.

It was a charmingly bright morning when we bade Hagerstown good bye, and took our places in the train on the Shenandoah Valley Railway bound southward. Passengers had come in on the Western Maryland Railway, and others on the Cumberland Valley, and now appeared after their breakfast at the station with smiling faces. Comparisons are odious, but a better meal than one gets at the railway restaurant in Hagerstown is unnecessary to either health or comfort.

"That's a point you're forever thinking about," says Prue, a little spitefully.

"I am, I acknowledge. It's of immense importance. Why is it I always prefer the Santa Fe route across the plains? Because I am sure of good meals. When one is traveling in the West or South, that consideration is doubly worth forethought. The certainty of finding well-cooked and abundant food was one great reason for my choosing this route for our present trip."

"Well, I wouldn't be so particular."

"Why not? It's largely your fault if I am."

"How, pray tell?"

"Because you have educated me to so good living at home!"

That softens the critic. Prue is justly proud of her tidy and accurate house-keeping.

The face of the country roughens somewhat south of Hagerstown, and a gradual but decided change in the appearance of things is noticeable. The special feature of the German farming region is preserved everywhere, however, north of the Potomac—I mean the huge barns. While the houses are generally comfortable and sometimes large, they are inconspicuous in the landscape beside the barns, which are magnificent—no simpler adjective will answer. They are not quite so big as Chicago elevators, but far more spacious than most churches. A few are built of wood upon a stone substructure which serves as a stable; but the majority are of stone with wooden sheds attached. The stone barns, having long slits of windows left for ventilation, resemble forts pierced for musketry; while a few new barns made of brick, secure the needful air by leaving holes, each the size of one brick, arranged in fantastic patterns up and down the gable ends.

The first stop out of Hagerstown is at St. James, a district full of reminiscence which Prue calls to mind at the sight of a group of buildings on the right a little beyond the station. This was "Ringgold's Manor," and Prue tells the story as we pass through the lands once under his sway.

Among the earliest settlers of this part of Maryland were the Ringgolds, whose estates amounted to 17,000 acres in one spot here, and much land elsewhere. The manor-house was at Fountain Rock, and was a splendid mansion decorated with stucco-work and carvings executed in good taste. "Many of the doors of the mansion," Prue recounts, "were of solid mahogany, and the outbuildings, appointments, etc., were of the handsomest character. The architect was the distinguished Benjamin H. Latrobe, who was also one of the architects of the national capitol at Washington. It was General Ringgold's practice to drive to Washington in his coach-and-four with outriders, and to bring this political associates home with him. Among his guests were President Monroe and Henry Clay. Mrs. Clay, you know," Prue adds, " was a Hagerstown girl named Lucretia Hartt. But this lavish hospitality and great extravagance finally worked Ringgold's ruin, and when he died his estate went to his creditors."

"Yes," Baily adds, "he had a jolly-dog way of lighting cigars with bank-notes, I have read; and each season would sell a farm to pay the expenses of the preceding congressional term."

The old manor-house was turned into St. James' College many years ago, but now only a grammar school occupies the premises.

The streams hereabout run in deep ravines and give good water-power. At Grimes station, the next stop, there is an old-time stone mill of huge proportions, with gambrel roof, exposed wheel and mossy flume, the whole surrounded by an orchard; near by stands the small, half ruined stone cottage of the miller, nearly hidden in the trees, making a charming subject for a picture.

Just beyond we get a small glimpse of a river, deep and powerful, seen down through a gorge which opens and shuts again as we leap its chasm. A few quaint houses (New Industry) fill the mouth of the gorge, but before we can look twice they are gone. Such is our first sight of the Potomac.

Not far eastward of Grimes is Sharpsburg and the mouth of the Antietam, a district which seems to have been especially populous in prehistoric days, and where an extraordinary number of relics and traces of Indian residence have been found. At Martinsburg lived a great settlement of Tuscaroras, and upon the Opéquon, which empties near there, dwelt a big band of Shawnees. At the mouth of the Antietam (which flows southward parallel with the railroad and two to four miles distant) there occurred in 1735 a memorable battle between the Catawbas and Delawares, for whom the Potomac was a border line, resulting in the defeat of the Delawares.

More thrilling war history than this makes this station memorable, however, for here, on September 17th, 1862, was fought a part of the great battle of the Antietam, the more central struggle of which took place in the plain eastward of the railway. Here at Grimes, however, was the extreme left of the Confederate line, where the trees are still scarred with the bullets, and the corn-fields conceal the wasted shot of that fatal day. A short distance beyond is a station called Antietam—the point of departure for Sharpsburg and its stone bridge, two miles distant, which lay at the heart of the hardest fighting. For three miles the railroad runs immediately in rear of the position held by the main command of "Stonewall" Jackson, and every acre of ground was stained by the blood of brave men. In the large brick house seen among the trees a short distance eastward of the station, General Lee had his head-quarters.

The United States soldiers' cemetery, where more than 5,000 of the Federal dead are buried, is near the village, but not in sight from the station; from the crest of the hill it covers, a general view of the whole battle-field can be obtained.

When we come upon the Potomac again it is with startling suddenness. Out of the clover and corn fields the train hides itself in a deep cut, and thence rushes forth upon the lofty bridge which spans the noble river at Shepherdstown.

Shepherdstown lies upon the southern bank and is one of the quaintest of villages. The cliff-like banks of the river are hung with verdure, few buildings skirt the water or nestle in the ravines which extend up to the level of the town, and on the northern side of the stream the famous old Chesapeake and Ohio canal still floats its cumbersome boats. At the head of a ravine stands one of those old stone mills, most temptingly placed for sketching, and the whole presentation of the town, with the green, still river curving grandly out of view beneath it, is one long to be remembered.

Having crossed the Potomac, we are now in the northeastern corner of West Virginia, and, in Shepherdstown, enter its oldest settlement, founded in 1734 by Thomas Shepherd, whose descendants still live there and own some of the original land. The pioneers were Germans from Pennsylvania chiefly, and the village has more the appearance of a Maryland than a Virginia town. Its settlement was followed closely by a large incoming of Quakers, who located themselves at the foot of the North mountain.

This community was active in revolutionary days, and from it sprang the first of those "buck-tail" mountaineers, who, recruiting as they went, hastened on foot to aid Washington, at Boston, in 1775, when he first called for troops. No incident in local history, however, is more important than the experimentation which was carried on here by James Ramsay, in 1785, toward the invention of a steamboat. The plan of the Chesapeake and Ohio canal was then under consideration, and projects for inland navigation were stimulating inventive thoughts. Washington and others became especially interested in what Mr. Ramsay was doing, and aided his experiments. Finally there was produced and tried on the Potomac a steamboat which unquestionably ante-dates the discoveries in this direction of Fulton and perhaps of Fitch. Ramsay's steamer was a flat-boat, "propelled by a steam engine working a vertical pump in the middle of the vessel, by which the water was drawn in at the bow, and

EVENING ON THE UPPER POTOMAC.

expelled through a horizontal trunk at the stern." The impact of this forcible stream against the static water of the river pushed the boat along, just as a cuttle-fish swims. This boat was eighty feet long, and, with a cargo of three tons, attained a speed up the current of four miles an hour. She was soon disabled, however, by the explosion of her boiler. Relics of her machinery are preserved in the National Museum, owing to the forethought of Colonel Boteler, of Shepherdstown.

During the late war Shepherdstown and its environs were the theatre of incessant army operations, and the town itself was shelled more than once by alternate guns. Its position made it an impracticable point for either army to hol 1, while its neighborhood was desirable to both. Hence in the evenly contested campaigns of the earlier years of the war, and the great marches and counter-marches which took place later, Shepherdstown was alternately occupied by both "enemies" to its peace and prosperity.

Walking in the evening to the high bluffs near the end of the fine bridge, and feasting our eyes on the beauty of the river-picture stretching away to Harper's Ferry, we can see, a mile below the town, ripples upon the water, near some large kilns and cement mills, which betokens a shallow place.

"There," I say to Prue, "is the famous old Pack-horse ford, which got its name in the colonial days when all the mountain paths were simply 'trails,' and the pack-horse the only means of transportation. Here would cross the northern savages when they went on their war expeditions against the southern tribes, and there emigrants and hunters and surveyors found their easiest transit of the river."

"I suppose," says Prue, "this must have been an important point in the late war, if, as you say, all the bridges were destroyed."

"It was. Soldiers were always crossing and re-crossing, but it became of especial use to Lee. By it a part of his army marched to the field of Antietam, and after the battle the whole of his forces re-crossed on the night of Septemper 18, to the Virginia side, at this ford. The main body of the Confederates continued their retreat inland, but a part of Jackson's army, under A. P. Hill, remained in partial concealment, and on that bluff which you see cleared just this side of the ford, batteries were planted. This was on the 19th of September, two days after the Antietam battle. Gen. Fitz John Porter, with the Federal fifth corps, had been ordered by McClellan to support the cavalry, and he determined to try to capture some of Hill's guns. He posted batteries on the knolls through which the railway passes at the northern end of the bridge, and lined the top of the Maryland bank with skirmishers and sharpshooters, supporting them by two divisions. Volunteers from the 4th Michigan, 118th Pennsylvania, and 18th and 22d Massachusetts regiments plunged into the ford at dark, and succeeded in capturing five guns. A reconnoisance in force was sent across the river next morning (20th), at seven o'clock. The cavalry ordered to co-operate failed to do so, and the unsupported infantry was sharply attacked by a greatly superior Rebel force. It was driven back, pushed over the cliffs, killed, captured, or forced into the river. The ford was filled with troops, for just at that moment the pet "Corn Exchange" regiment of Philadelphia was crossing. Into these half-submerged, disorganized and crowding masses of men, were poured not only the murderous fire of the Rebel cannon and rifles, but volley after volley from the Federal guns behind them in trying to get the range of the Confederate batteries. The slaughter was terrific. The Potomac was reddened with blood and filled with corpses. When the routed detachment struggled back to shelter, a fourth of the Philadelphians, who had been in service only three weeks, were missing, and their comrades had suffered equally.

Thus week after week, and year after year, did Shepherdstown and the lower part of the Shenandoah valley hear the thunders and witness the devastation of war.

VI.

THE LOWER VALLEY OF THE SHENANDOAH.

Shenandoah Junction with B. & O. R. R.—Charlestown and "John Brown's Body"
—Harewood House.—Approaching the Blue Ridge.—Berryville and its
New Railroad to Washington.—Recollections of the Early and
Sheridan Campaign.—The Old Chapel.—The Home of
Lord Fairfax.—First Sight of the Shenandoah.
—Front Royal and its Fights.—
The Massanutten.

A FEW miles above Shepherdstown the track crosses (upon a bridge) the main line of the Baltimore and Ohio Railroad. The station is called Shenandoah Junction, and here passengers change cars for the West and for Washington. Near this point lived a trio of officers in the Revolutionary war whose histories were sadly similar—Horatio Gates, Charles Lee and Adam Stephen. All were with Washington at Braddock's defeat and all were there wounded; all became general officers in the Continental army; and, finally, all three were court-martialed for misconduct on the field, and found guilty.

Before we have fully recalled these facts to each other, we cross another railway—that from Harper's Ferry to Winchester, which was so useful to Sheridan—and are at Charlestown, a place marked chiefly in my recollection as the former home of that talented and lamented humorist "Porte Crayon." The village lies off at the left of the track, behind a square mile or so of corn fields, and is a thriving town of about 2,500 people. It is built upon lands formerly owned by Charles Washington, a younger brother of the general, and was named after him.

Lying upon the direct course between the river-gap at Harper's Ferry (Loudon Heights rear their noble proportions just behind the town) and the principal villages of the valley, Charlestown has had its share in all the principal episodes of the history of the region. Hither came Braddock's boastful army and a well is pointed out, close to the railway station, which was dug by them. Hither, too, was brought John Brown—"Brown of Ossawatomie"—to be hanged, and you may see a great number of relics connected with his career. The court house in which he was tried and the field where he was executed, are both visible from the cars.

This way, too, following the standard held aloft as "his soul went marching on," came the first Union troops that entered the Valley of Virginia, and every by-road here was the scene of continual fighting, beginning with the "demonstration" made by Jackson immediately after the battle of Winchester. Later Sheridan and Early sparred at each other over this ground, Early having great success at first, but finally compelled to relinquish what he had gained.

"Why it must be near here," says Prue, as we are moving off "that Harewood House stood."

"It stands only a mile or so toward the west, and not far away you might find the remarkable ruins of a stone church, erected during the reign of George II."

"What was 'Harewood House'?" Baily inquires.

"The home of George Washington's elder brother Samuel," he is informed. "It was built under the superintendence of Washington himself, and still stands unchanged—a valuable example of the architecture of its time."

"Ah," Prue adds, "that house has seen some fine times and fine people! James Madison was married in it; and there Louis Phillipe and his two ducal brothers, Montpensier and Beaujelaix, were entertained as became princes."

"A DOT OF A CABIN."

The face of the country waxes hilly as we proceed, and at Fairfield we find ourselves close to the foot of the Blue Ridge. It is no longer hazy blue, but green; its features are distinctly visible, and here and there a dot of a cabin appears, but no large clearing anywhere. The great Dutch barns have disappeared, and the broad square faces of the Dutchmen are exchanged for the thin countenances of the Virginians. Every notch through the mountains has its name, first Yeskel, then Gregory, then Rock, then Snicker's. The last, though abruptly walled and picturesque, will admit the passage of a railway, and through it is now being built the extension of the Washington, Ohio and Western, finished as far as Round Hill. This road, proceeding westward across Loudon county, the old home and retreat of Mosby's guerillas, and worming its way through Snicker's gap, will join the Shenandoah Valley's track at Berryville, and soon form an independent, shorter and highly attractive route between the South and Washington.

All along on our right, the ground was somewhat higher than where the tracks ran, yet not high enough to impede the view of the regular front of the Little North mountain, here about twelve miles directly westward. This slight elevation is called Limestone ridge. It runs lengthwise of the valley, and the rainfall upon its opposite slope drains into the Opequon (O-pe'k-on).

Eleven miles above Charlestown is Berryville, the county seat of Clarke, which has been called the "most interesting county in the valley to the student of history." The place owes its importance to the fact that it lies upon one of the great thoroughfares over the Blue Ridge—the turnpike through Snicker's gap—and to the fertile country by which it is surrounded. Berryville will begin a second prosperity when the new railroad I have mentioned is completed from Washington to this point. Prue asks why, long ago, it was called "Battletown," and I cannot tell her; but there has been abundant reason since for such a name. Banks took possession of the place as early as '61, following the macadamized road from Harper's Ferry to Winchester. In 1864, when Early was retreating from his Maryland campaign, loaded with plunder, here occurred a sharp fight; subsequently Sheridan made this point a centre of extensive operations; and on September 3, 1864, by a mutual surprise, a battle was precipitated in the afternoon between a large Confederate force and the Federal eighth corps, which ceased only when it was too dark to see. By the way of this turnpike, too, were sent forward the great armies that pressed back Early's forces after the battles around Winchester.

A little way past Berryville Prue calls us hastily to look down at the right upon an old cemetery crowded with headstones, and shaded by a growth of aged trees beneath which the tangled roses and untrimmed borders of redolent box have flourished unchecked. A stream, mourned over by weeping willows, creeps stealthily by; and in the midst of the graves stands an antique chapel approached by several roads.

"Is it not peaceful and comforting?" cries Prue. "I think one might lay a friend in such a place as that with 'sweet surcease of sorrow' far different from the bleak repulsiveness of most rural cemeteries."

"Yes, and that, perhaps, is the feeling with which at a certain time every year the old families whose country-seats have been in this region for many generations, assemble for a day of memorial services over their dead who are buried under those stately trees."

" I am told," Baily adds, "that its first pastor was Bishop Meade, the same who wrote a book upon the old churches and old families of Virginia, which contains the full history of this chapel."

VIRGINIA UPLANDS.

The locality into which we are now so swiftly and smoothly penetrating is one replete with landmarks and traditions of Colonial history. A mile or two beyond Boyceville, for instance, we observe, off at the right, a stone house of old-fashioned style, which has been known for a century as "Saratoga," because built by Hessian prisoners captured with Burgoyne.

Then comes White Post.

"Strange name for a station," Prue remarks. "How did it arise?"

"This," I say, "was the centre of that great estate, of more than five millions of acres, granted by the English crown to Lord Fairfax, Baron of Cameron, the boundaries of which included all the region between the Rappahannock and the Potomac, marked on the west by a line drawn from the head springs of the one to that of the other river. It was the task of the youthful George Washington to survey that part of this vast estate beyond the Blue Ridge, and it was in pursuance of this duty that he made the western trips and tramped over the country in the adventurous way we have read about. Near the intersection of the roads from the two main gaps through this part of the mountains, Lord Fairfax built himself a country house of no great size or elegance; and at the junction of the roads he set up a white oak finger-post as a guide. The original post still remains, carefully encased for preservation."

" Is the house still standing?"

" No ; but there is a new one on its site. Fairfax called it 'Greenway Court,' and with the open, lavish hospitality characteristic of rich frontiersmen, he made it the scene of revelry and rough, hilarious sports, such as were enjoyed by the carousing, fox-hunting generation in which he lived. It was his intention to have erected a larger and more pretentious mansion, but this project was never carried out, and the proprietor lived the remainder of his days in the house first erected. Here he dwelt when his former protege, Washington, had successfully prosecuted the war for independence to the surrender of Cornwallis at Yorktown, and the deliverance of the colonies had been achieved. Strongly attached to the English cause, when told of the surrender he turned to his faithful servant and remarked : ' Take me to bed, Joe; it is time for me to die.' Old and feeble at the time, he never rallied, dying December 9, 1781."

Here Prue points out a noble height coming into view directly ahead, which seems to lie right in the centre of the valley.

"That," she is informed, " is Massanutten mountain, or The Massinetto, as it is given in early writings."

"Yes," Baily interposes, "and here, at last, is the Shenandoah, the beautiful stream that with keen poetic instinct the Children of the Forest named *The Daughter of*—"

"That will do, Baily ; you don't know anything about it."

"Well, if it don't mean that, what does the name signify ? "

"Nobody seems to know, at any rate, *you* don't. Why, its very spelling is so obscure that probably we have lost the original word entirely. In the earliest accounts it was the 'Gerando,' then the 'Sherando,' or 'Sherandoah,' and the present spelling is quite recent."

"Anyhow, here's the river ! "

"Yes, and isn't it a beautiful one ! " Prue exclaims. " I have heard a traveler say that ' it deserves the epithet *arrowy* as well as the Rhone.' Surely, it should have a poetical name."

"And *has* a *musical* one, which is much more to the purpose," I insist. "See how graceful are its curves, how silken and green its quiet current, how deeply embowered in foliage and rocky walls, and what pretty little gateways are broken down through them to let the hill-brooks pour their contributions into its steady flood !"

A few moments later we cross on an iron bridge at Riverton, the point of confluence of its two forks—the "North" and the "South." The North fork comes down from the other side, and its basin is distinguished as the Shenandoah valley proper, while our route lies between the Massanutten and the Blue Ridge, that is, up the South fork. This is generally spoken of simply as South river, and its basin is called the Page valley. At Riverton the Manassas branch of the Virginia Midland Railway (which figured so largely in army movements during the civil war) crosses *en route* from Manassas to Strasburg, and there are evidences of an important manufacture of lime. The village itself is out of sight, as also, is Front Royal, whose station is called two miles ahead. I told Baily to stop and go over there, while Prue and I went on to Luray; and his report was so glowing I regretted we had not been with him.

To the site of Front Royal, according to Baily, came white settlers as early as 1734, and placed their houses in a sheltered nook among the hills beside the Shenandoah, at a point where the Indian trails from Manassas and Chester's gaps joined into one near the mouth of a little stream since called Happy creek. This fact produced a Y-shaped settlement, which, with the increasing growth of the village, has not been changed, the three main streets still following the old paths marked out by the moccasined feet of pre-historic pedestrians. Gradually the fame of the fertility and beauty of the Valley of Virginia attracted settlers from the coast and from abroad, and the Indians were replaced by hardy white men. This new settlement, then called Lehewtown, became a centre of a large district and attracted so many rough characters that

BETWEEN FRONT ROYAL AND LURAY.

it came to be known as "Helltown," with good reason. By the close of the Revolution, however, order and respectability prevailed, and in 1788 a town was incorporated, under the name of Front Royal, the origin of which term is a nut for historians to crack. From that time on it has been prosperous, having acquired wealth and fame in manufactures as well as through its rich environment of farms and vineyards. There were made the celebrated Virginia wagons of a past day, which were the best of their kind in the whole country, and were taken by emigrants to every new state and territory as forerunners of the prairie schooner. Hand-made and durable as the "deacon's one-hoss shay," their cost was so great that the machine-made wagons have surpassed them as thoroughly as the cradle has overcome the sickle; but Front Royal still shapes and sells great quantities of spokes, hubs and other wagon material of the best quality.

Front Royal is now a neat and pretty village, of perhaps a thousand people, which is growing rapidly. As the county seat of Warren it becomes the residence of the professional men of the district, and is marked by a society of unusual intelligence.

Here occurred some exceedingly interesting incidents during the war, in one of which a mere handful of Confederate cavalry under a boyish commander dashed into the village, captured the provost guard, and made off with it successfully, though two whole regiments of bewildered Federals were at hand to protect the place. Ashby (whose birthplace and home was up in the Blue Ridge, not far away) was hovering about here much of the time, while Jackson enacted his series of victories in this district ; and on May 22, 1864, here took place one of the most disgraceful routes Union soldiers ever were ashamed of, four companies of Flournoy's Virginians attacking a thousand or so of Banks' army, entrenched on Guard hill, with such impetuosity as to scare them in utter confusion from their works, with great loss of life, stores and artillery. These disasters were requited later in the same year, however, when Sheridan, driving back Early, fought so stubbornly along this very limestone ridge which the railway track follows ; and Front Royal echoed again and again, during that and the subsequent year, to the roar of cannon, the sharper crackle of small arms and the hoofs of charging cavalry.

From Front Royal station southward to Luray the line passes through a region of wooded hills and deep ravines, the latter crossed upon some very high trestles. The river is often close beneath the track, and its course through these rocky highlands presents many views that excite our admiration. We are fairly among the foot-hills of the Blue Ridge here, though its central peaks are far enough away to show to good advantage. In this rough district, where more wooded than cleared land is seen, a fine grade of "neutral hematite" iron ore occurs, the principal point of shipment for which is at Rileyville. A few momen's after leaving that station we are at Luray, and have alighted to take a pleasant night's rest, and see the wonderful caverns.

VII.
LURAY AND ITS CAVERNS.

Old Caves.—Discovery of the New Caverns.—Startling Effect of Electric Light in the Cave.—Theory of Excavation.—A Rapid Survey.—The Bridal Chamber and its "Idiots"—Varieties of Stalactite.—Richness of Color.—Musical Resonance.—The Skeleton.—A Fair World.—Value of a Good Hotel.—Luray as a Summer Residence.

PAGE valley is here several miles wide, and the surface is diversified by an endless series of knolls, ridges, and deeply embedded streams. "The rocks throughout the whole of this region have been much displaced, having been flexed into great folds, the direction of which coincides with that of the Appalachian mountain-chain. In fact these folds are a remnant of the results of that series of movements in which the whole system primarily originated." Hidden in the woods near the top of one of these hills, about a mile east of Luray, an old cave has always been known to exist. Connected with it are traditions which reach back to the Ruffners, the earliest settlers of the valley, and it has taken their name.

In 1878 Mr. B. P. Stebbins, of Luray, conceived the project of a more complete exploration of it, with a view of making it an object of interest to tourists, and he invited the co-operation of the brothers Andrew and William E. Campbell. These gentlemen declined to go into the old cave, but were ready to engage in a search for a new one, and went ranging over the hills, but for four weeks succeeded only in exciting the astonishment and ridicule of the neighborhood, when, returning one August day from a long tramp, the men approached home over the hill where Ruffner's cave was. In the cleared land on the northern slope, a couple of hundred yards or so from the mouth of the old cave, was a sink-hole choked with weeds, bushes, and an accumulation of sticks and loose stones, through which they fancied they felt cool currents of air sifting.

Laboriously tumbling out the bowlders, Mr. Andrew Campbell was finally able to descend by the aid of a rope into a black abyss, which was not bottomless, however, for he soon let go of the rope and left his companions on the surface to their conjectures. Becoming uneasy at his long absence, his brother also descended, and together the men walked in a lofty passage for several rods, where their progress was stopped by water. Returning, they told Mr. Stebbins what they had seen, and all agreed upon a policy of silence until the property could be bought. Then they went home and dreamed of "millions in it." Such was the discovery of the Luray cave.

Dreams are but a "baseless fabric." The property was bought of a bankrupted owner, at sheriff's sale, but upon an intimation of its underground value, one of the relatives of the original owner sued for recovery upon an irregularity in the sale, and after two years of tedious litigation, he won his suit. Previously a company of Northern men, of

HALL OF THE GIANTS.

whom Mr. R. R. Corson, of Philadelphia, is president, had formed a joint-stock company to purchase the property, and it passed into their hands in the spring of 1881. But during the two years the original cost had swelled, while the early visions had dwindled, until they met at $40,000. This is the history of the "wonder," and now we are ready to enter it.

But it is over a mile from the hotel to the cave, and the day is warm. Enquiry develops the information that if we are willing to wait until some train arrives we may find hacks at the station which will take us the round trip for thirty-five cents; but if we wish to go at our own convenience the clerk at the hotel will summon a hack when we please, and we must pay fifty cents fare.

I had been intending to buy Baily a certain cigarette-holder which had taken his fancy, as a present, but I reflected that if instead I took the money it would cost and applied it to paying the extra charge of the latter alternative we would enjoy the trip better, so I told Mr. Mullin that I would ask him to telephone for a hack at once. This was after breakfast on the morning succeeding our arrival.

"What shall I wear?" asks Prue, "I suppose it's a horrible muddy and soiling place. I shall envelop myself in my waterproof, of course, but how about a bonnet?"

"No change whatever is needful," we were told. "You will find an even temperature of about 56° Fahrenheit throughout the cave, and all the year round. There is plenty of room to walk about everywhere without squeezing against the walls or striking your head, and board or cement walks and stairways are provided throughout all the area open to visitors. It is advisable, nevertheless, for ladies to wear rubbers, since there is enough dampness underfoot in some places to penetrate thin-soled boots."

So Prue resumed her traveling dress—that short-skirted, close-fitting wine-colored flannel I like so much—donned a snug turban and off we went. I told Baily he'd better leave his crutch-headed cane at home, but he is a bit of a dandy and wanted to "show it to the natives," so I had the laugh on him when it was taken from him by the keepers of the cavern, who wisely allow no dangerous implements of that kind among the fragile treasures of their underground museum.

Our road led us through the long main street of the village, but we attracted little or no attention, for nearly 20,000 tourists a year ride up and down this stony street. Half a mile beyond the town, on the slope of a low hill, stands a house with porticoes all around it and a public air. Here we registered our names, paid our admission fee, and were assigned to the charge of a guide. His first act was to hand to me a sort of scoop-shovel reflector, or sconce, in which were placed three lighted candles, and take another himself. This made us look at one another, as much as to say—"This thing is a humbug!" for we had been told of far better means of illumination than that; but meanwhile the guide had opened an inner door and invited us to follow him down a staircase of masonry, and, before we supposed our day's adventures had begun, we found ourselves in the large antechamber of the caverns. This unpremeditated, unintentional entrance is as though you had been dropped into the midst of it, or had waked from a sleep there, and is most effectual in putting the stranger *en rapport* with the spirit of the place.

The darkness was only faintly illuminated by our few candles, and I was about to remonstrate, when the click and flash of an electric arc, flooded the whole place with light. Our few candles were intended merely for peering into dark corners and helping our footsteps—the general illumination is accomplished by dozens of electric lamps hung in all parts of the wide-winding vaults and passages. As soon as I perceived this I gave my sconce to Baily, for it was a nuisance to carry it.

This first chamber is about as big as a barn (*not*, a Cumberland valley barn, Prue wisely remarks), and from it we proceed upon a causeway of cement for a short distance past the Vegetable Garden, the Bear Scratches, the Theatre, the Gallery ; over Muddy Lake on a planking bridge, which is itself spanned by a stone arch ; through the Fish Market and across the Elfin Ramble—a plateau in which the roof is generally within reach of the hand ; and so come to Pluto's Chasm.

Gazing down over the edge of this underground ravine, Baily exclaimed: "What mighty convulsions must they have been which rent these walls asunder!"

"There, Baily, is where you show your—well, your insufficiency of knowledge ! This chasm owes its configuration to the same slow and subtle agencies that produce a cañon above ground in this limestone valley."

"Why do you say 'limestone' valley?" Prue asks.

"Because great caves can only occur in a limestone region, since they result from the chemical fact that carbonates of lime and magnesia are soluble in water containing carbonic acid. This acid abounds in atmospheric air, and is one of the products of the decomposition of animal and vegetable waters, so that rain-water which has percolated through the soil has usually been enriched with it from both sources. Let this chemically charged water find its way into some crevice, and it only requires time and abundance of water to dissolve and hollow out Pluto's and all the other chasms, halls, galleries and avenues ; and when once this work has well begun, other natural agencies contribute their aid to the enlargement of the area and the adornment of its interior.

From the chasm, where there is a Bridge of Sighs, a Balcony, a Spectre, and various other names and habitations, we re-cross the Elfin Ramble, pass successively Titania's Veil, Diana's Bath—the lady was not fastidious—and come to a very satisfactory Saracen Tent.

Then we ascend stairways past the Empress Column—easliy empress of all, I think—and proceed under the Fallen Column to the spacious nave of the Cathedral. We pause to note its lofty groined roof and Gothic pillars—surely, in some like scene to this, the first architect of that style met his inspiration !—its large, Michael-Angelesque Angel's Wing, and its Organ. Then we sit down and turn to the prostrate stalactite. It is as big as a steamboat boiler, and bears an enormous pagoda of stalagmitic rock which has grown there since it fell. It thus forms a good text for a conversation, as to the age and geology of the cave, the materials for which we found by reading an excellent pamphlet on the subject published by the Smithsonian, and

BANKS OF THE RHINE, "LURAY CAVERNS."

which may be procured at Luray. The gist of it is, that the cave is probably considerably later in its origin than the close of the carboniferous period, and not more ancient than the Mammoth or Wyandotte caves. The indications are, that in past ages the work went on with great rapidity, but that latterly change has been very slow, and at present has almost ceased.

Leaving the Cathedral, a narrow, jagged passage, we get an outlook down into a sort of devil's pantheon, full of grotesque shapes and colossal caricatures of things animate and inanimate, casting odd and suggestive shadows in whose gloom fancy may work marvels of unworldly effect, and then are led by a stairway to a well-curtained room called the Bridal Chamber.

"Was anyone ever really married here?" asks Prue, incredulously.

"Three couples, so far, Madame," the guide informs her.

"Well!" exclaims the neat little lady. "I had no idea there were such idiots! Now if you had said three funerals, I could have found some appropriateness in it."

The back door of the Bridal Chamber admits to Giant's Hall, just beyond which is the Ballroom—both large and lofty apartments, constituting a separate portion of the cave, parallel with the length of Pluto's Chasm. In the Ballroom we have worked back opposite the entrance, having followed a course roughly outlined by the letter U.

I have thus run hastily over the greater part of the ground open to the public, in order to give an idea of its extent and nomenclature. To describe each figure and room separately is impossible. The best I can do is to try to give some general notion of the character of the ornamental formations of crystalline rock which render this cave without a peer in the world, perhaps, for the startling beauty and astonishing variety of its interior.

Though the simple stalactite will be circular and gradually decreasing in size, conically, from its attachment to its acuminate point, yet innumerable variations may occur, as the dripping or streaming water that feeds it is diverted from its direct and moderate flowing.

Chief of all the varieties, and the one that in lavish profusion is to be seen everywhere in these caverns, is that which, by growing on the edges only, produces not a round, icicle form, but a wide and thin laminated or sheet form, which is best described by its semblance to heavy cloth hanging in pointed folds and wrinkles, as a table-cover arranges itself about a corner. Where ledges and table-like surfaces—of which there are many instances in the cave—are most abundant, there the "drapery" is sure to form. In the Market it crowds the terraced walls in short, thick, whitish fringes, like so many fishes hung up by the gills. The Saracen Tent is formed by these great, flat, sharply tipped and gently curving plates, rich brown in color, depending from a square canopy so that they reach the floor, save on one side, where you may enter as through conveniently parted canvas. The Bridal Chamber is curtained from curious gaze by their massive and carelessly graceful folds ; the walls of Pluto's Chasm are hung with them as in a mighty wardrobe ; Diana's Bath is concealed under their protecting shelter ; Titania's Veil is only a more delicate texture of the same ; Cinderella Leaving the Ball becomes lost in their folds as she glides, lace-white, to her disrobing ; and a Sleeping Beauty has wrapped these abundant blankets about her motionless form ; while the Ballroom carries you back to the days of the Round Table, for the spacious walls are hung as with tapestries.

Do not disbelieve me when I speak of wealth of color. The range is small, to be sure, but the variation of tint shade is infinite and never out of tune. Where the growth is steady and rapid, the rock is crystal white as at the various Frozen Cascades, the Geyser and many instances of isolated stalactites. But when the steady growth ceases, the carbonic moisture of the air eats away the glistening particles of lime, and leaves behind a discolored residuum of clay-dust and iron oxides. Thus it happens that, from the niveous purity or pearly surface of the new work there runs a gentle gradation through every stage of yellowish and whitish brown to the dun of the long abandoned and dirty stalagmite, the leaden gray of the native limestone, or the inky shadow that lurks behind. It is thus that the draped and folded tapestries in the Ballroom are variegated and resplendent in a thousand hues. Moreover, various tints are often combined in the same object, particularly in the way of stripes more or less horizontal, due to the varying amount of iron, silica, or other foreign matter which the lime-water contained from time to time.

The best example of this, and, indeed, of the "drapery formation" generally, is to be found in the Wet Blanket. A large number of the pillars are probably hollow, and are formed by the crowding together of many drapery-stalactites, which finally have coalesced, leaving the pillar deeply fluted, or seamed up and down, along their connected edges. When you find one of these massive, ribbed and rugged pillars vanishing above in a host of curved stalactites, their thin and wavy selvages guiding the eye to tips which seem to sway and quiver over-

A MOUNTAIN CASCADE.

head, it is hard not to believe it is an aged willow turned to stone, Indeed the whole scene, in many parts, is strongly suggestive of a forest with tangled undergrowths, thrifty saplings, fallen logs, and crowding ranks of sturdy trees.

In more than the general effect, indeed, the ornamental incrustations of this cave mimic the vegetable growths outside. Many of the stalactites are embroidered with small excrescences and complicated clusters of protruding and twisted points and flakes, much like leaves, buds, and twigs. To these have been given the scientific name of

LURAY INN.

helictites, and the grottoes of Stebbins Avenue exhibit them to the best advantage.

Then there are the botryoids—round and oblong tubers covered with twigs and tubercles, such as that cauliflower-like group which gives the name to the Vegetable Garden ; these grow where there is a continual spattering going on. A process of decomposition, dissolving out a part and leaving a spongy framework behind, furnishes to many other districts quantities of plant-semblances, that you may name and name in endless distinction. Then, in the many little hollow basins or "baths," and in the bottom of the gorges where still water lies, so crystal clear you cannot find its surface nor estimate its depth—where

STATION AND RESTAURANT AT LURAY.

the blue electric flame opens a wonderful new cave beneath your feet in the unrecognized reflection of the fretted roof, and where no ice is needed to cool nor cordial competent to benefit the taste of the beverage—there the hard gray rock blossoms forth into multitudes of exquisite flowers of crystallization, with petals rosy, fawn-colored and white, that apparently a breath would wilt.

But I must cease this attempt at even a suggestion of the possible variety of size and shape, mimicry and quaint device to be met with in this cavern.

That rigid stone should lend itself to so many delicate, graceful, airy shapes and attitudes, rivaling the flexible flower of the organic world, fills the mind with astonishment and bewilders the eye. And when you have struck the thin and pendent curtains, or the "pipes" of the Organ in the Cathedral, and have found that each has a rich, deep, musical reasonance of varying pitch, then your admiration is complete. The impression of it all made upon such visitors as are affected at all beyond ohs! and ahs! if written down, would form very curious reading; but little has been recorded, chiefly because it is one of the most difficult things in the wide world to do adequately.

The cave has not yet much human interest; but we must not forget to follow down a long stairway into a deep and narrow gluch, where the dampness and gloom is little relieved by anything to please the eye. At the foot of the staircase the guide drops his lantern close to a trench-like depression, through which a filmy brooklet trickles noiselessly. No need of interrogation—there is no mistaking that slender, slightly curved, brown object, lying there half out, half embedded in the rock, with its rounded and bi-lobed head, nor its grooved and broken companions. They are not fallen, small stalactites; they are human bones. Fit for the mausoleum of emporers, what a vast vault to become the sarcophagus of one poor frame!

Out into the warm, sweet air again, all the world looks fairer for one's temporary occultation. Surely the Troglodytes had a hard lot. Even the Naiads under the water, and the Dryads, though indissoluble from growing trees, were better off!

And what a fair world it is! How prodigal of beauty are soil and sun! How grandly has the architect and landscape-gardener of the globe adorned this valley! How precious the scene to him whose beloved home is here; and how novel and entertaining its features to the stranger!

Rested and well-fed we sit upon the piazza of the inn and thank the good fortune which brought us hither. No one can appreciate a good hotel better than he whose ill-luck it was to travel in the South a dozen years ago, where that article was unknown. The people who owned and prepared the cave, and the railwaymen who meant to profit by it, knew that the country taverns would never do. They built on this hill top, in the midst of a populous valley which was not only pleasant to look at, and charming to ride and walk over, but which could supply all the fresh vegetables and fruit and meat so desirable upon a rural table; a hotel constructed after that most picturesque design—the Early English—and including all the modern appliances for health and comfort. Beyond the ornamental grounds, we see puffs of steam coming from a half-hidden building. There is where the water is pumped up to the hotel, where the gas is made which illuminates all its rooms, and where the dynamo is placed which supplies the electric lights of the cave through a circuit over seven miles in length.

The Luray Inn, then, is not only a charming stopping-place for the casual transient tourist who stops off only half a day to see the caves, but offers an attractive residence to visitors who may choose to stay a week or a month or a whole summer.

No part of the valley is more interesting. If historically disposed, the visitor may reconstruct the odd life which went on here a century and a half ago, whose quaint customs are not yet forgotten.

"Who were the settlers here at first," Prue inquires, "and what does this queer name Luray mean?"

"One question answers the other. This part of the valley was settled first by Huguenots who had escaped from France thro' the Palatinate; and they named their district Lorraine, which has been corrupted into *Luray*, by changes really slight when you think of the eliptical tendency of all pronunciation in Virginia."

If recent history is more attractive, then here is the place to gather thrilling reminiscences of the long campaigns of the civil war from Jackson in '61 to Sheridan's victory in '65, which belong to every hill-

AN "INTERIOR" IN THE INN AT LURAY.

top and each valley road. If one enjoys sport, here are the forests and stream of the Blue Ridge or the Massanutten, and

> "——a full-fed river winding slow,
> By herds upon an endless plain."

If he is an artist—surely he could find no richer field. Luray itself is a relic of the old-time Virginia rural villages—quaint, irregular, vine-grown and full of romantic suggestion. Along the river, pictures of the most enchanting character may be found; with the water in the foreground, a rocky wall right or left, a middle distance of farm-lands and well-rounded copses, the vista will always lead straight to the clustered peaks that stand proud and shapely on the horizon.

> " Ah, what a depth in that blue sky,
> With rugged mountains softly blent;
> As here we wandered, you and I,
> Singing, painting, as we went."

VIII.
UP THE SOUTH FORK.

The Hawksbill.—Shield's Pursuit of Jackson.—"Stonewall's" Personal Fighting.—Elkton.—The Battle of Port Republic.—Iron Mining and Manufacture.—Other Minerals.—The Way to the White Sulphur Springs.—Jubal Early's Defeat at Waynesboro.

FROM Luray southward the road runs upon a ridge separating the Shenandoah from the Hawksbill, which was crossed on a high trestle just at town, and whose broad valley is filled with prosperous farms. It was a favorite resort for cavalrymen during the late war, since they not only found it a capital region to operate in, but plentifully stored with forage. Through the many passes in this part of the Blue Ridge would descend the troopers of Mosby, and to the same fastnesses fled the horsemen of Early's hard-pressed squadrons, only to re-appear again the moment the coast was clear.

Up this South fork, in 1862, Shields hastened forward after Jackson, who had escaped between him and Fremont at Strasburg, while the latter commander chased him up the North fork. The plan was to unite at the southern end of the Massanutten, and there defeat the weary and weakened Rebels by means of their combined forces—a plan which promised success, but failed to keep its promise.

Shields' first care was the bridges, of which three spanned the Shenandoah between Luray and Port Republic. One of these was just here opposite Marksville station (a place now noteworthy for the superior ochre which is mined in its vicinity), but he was too late, for Jackson had burned it. Thus compelled to take muddy roads (this was the first week of June), he struggled slowly along the western bank of the river until his advance had arrived at Conrad's store, where was the next bridge, and which is only a mile or two from our station, Elkton, on Elk run.

(It was by the way of Swift Run gap and down this little side valley, that Spottswood and his "Knights of the Golden Horseshoe" first looked upon the Shenandoah, in 1716, whence sprang the Scotch-Irish ancestors of the land-holders of this region.) Carroll, one of Shield's subordinates, pushing north to secure the bridge at Conrad's, with Tyler's brigade a few miles behind, surprised the whole of Jackson's trains and camp, left under the guard of only a few cavalrymen with three guns. Dashing in, Carroll nearly stampeded the train and escort, but it happened that the commander and his staff were there, and taking part himself in the very front of the skirmish, Jackson succeeded in recapturing the bridge, beating back the bold Federal squad, and recovering his equipage. Meanwhile the battle of Cross Keys, a few miles to the westward had begun, Ewell's Confederates facing Fremont and holding him in check until night allowed the vanquished Federals to retreat.

All this time our merry train has been carrying us southward, and when the whistle sounds for Port Republic—the next station above Elkton—we are running straight across the river-plain on which was fought the frightful battle of June 10th, 1862, where the dead lay so thickly that Jackson thought they must outnumber the living.

Here is the head of the South fork of the Shenandoah, and the town takes its name from the fact that formerly flat-boat navigation began at this "port." About four miles southwest, the North and Middle rivers, the principal tributaries that go to make the main Shenandoah, unite, and at this point, South river, coming from the base of the Blue Ridge, joins them. In the angle between South and Middle rivers lies the town, and through it goes the valley turnpike on its way to the crossing of the Blue Ridge at Brown's gap. From the cultivated river-plain a succession of terraces arise to the wooded spurs of the mountains.

On the morning of the 10th of June, the Union army under Shields had been planted below the town in a very advantageous position. Jackson's men were divided, but withdrawing Ewell's army from its position at Cross Keys, Jackson soon outnumbered the force of Shields, who could expect no help from Fremont. The fighting began early in the day and was especially severe in the elevated woods upon the left of the line of battle, where Tyler's Federal guns were captured and re-captured by hand-to-hand conflicts in the thickets. At first the Confederates got the worst of it, and their general trembled for the result; but his arrangements were so careful, his celerity in re-inforcing was so great, and his men were so recklessly courageous, that they bloodily snatched victory from defeat and pressed the Federals so heavily that for a short time the retreat became a rout. The loss was terrific—a far larger percentage than is usual in battles; and though the cavalry began to follow the fleeing foe they were speedily recalled, and before night the whole Confederate army was hastily withdrawing into the security of Brown's gap, Fremont, who had come to the bluffs on the western bank of the river giving them a parting salvo.

Meanwhile Shields (and later, Fremont), under orders from

McDowell, continued the retreat to the base of operations in the lower valley. These battles of Cross Keys and Port Republic closed Jackson's momentous and brilliant campaigns of 1862—closed them in the very region where they were begun with a small and dispirited army only three months before. The succeeding week he spread his camps in the park-like groves and dells which lie a little south of Port Republic—the very hills through which the track now winds so ingeniously.

But*Baily, who has a practical turn of mind far above me, has been listening to only a portion of my war stories, having gone off to chat with a gentleman whom he somehow discovered was informed about iron matters in these hills. Reporting this conversation, Baily tells us that this region is full of metallic wealth and has long furnished iron and various other useful minerals to commerce, rivaling the mining districts of the Appalachian ranges north of the great valley. On the Massanutten outcrops of iron ores, classified as Clinton Nos. III and V, occur in nearly every peak, while universally, almost, at the western base of the Blue Ridge, primordial iron comes to the surface.

"We have just passed," says Baily, "at the station called Milnes, between Luray and Elkton, the large Shenandoah Iron Works, where for many years charcoal iron has been made, but now blast furnaces have been erected and coke-iron is made. They tell me that the company owns 35,000 acres of land along the foot of the mountains, only a small portion of which is under cultivation, and that the iron ore is quarried out of open excavations."

"What sort of iron is made?"

"The ore is a brown hematite, and the product is a neutral iron of especial value for foundry use. Only pig is cast now, but blooms can be made when the market justifies it. About three hundred and fifty men are employed."

"Is that all there is at the station?" I ask.

"No, it is the end of a division of the railway—you noticed that we changed locomotives; and there are small repair shops. The result is a busy little town which furnishes the neighboring farmers so steady a market for their beef, poultry, garden produce and forage, that they are well-off and enhancing the value of their lands by steady improvements and a higher style of agriculture. Sixty or seventy dollars an acre is asked for the best farms in that neighborhood, though a great deal of unimproved land may be bought for ten dollars an acre."

Iron, however, is not the whole mineral wealth of this region. Umber, ochre, copper, manganese, marble, kaolin, fire-clay and various other useful metals and earths are known to lie adjacent to the line of railway we are following, and are rapidly being availed of by capitalists. A complete account of these resources has recently appeared in a compact volume written by Prof. A. S. McCreath, while, in the files of that admirable monthly, *The Virginias*, published by Major Jed. Hotchkiss, at Staunton, detailed information and statistics may be found.

"By the way," Baily remarks, as the train pulls up at Waynesboro

Junction, a mile from the large and well known town of Waynesboro, "Hotchkiss says this place deserves a name of its own, because it is going to be a great town some day."

"Why does he think so?"

"On account of the ease of transportation to it from four directions of the crude materials; of minerals and timber property abounding in the

THE BLUE BRIDGE, NEAR WAYNESBORO.

region to which it forms the centre, and of the machinery necessary to their manufacture."

Just now Waynesboro is merely the crossing of our road by the Chesapeake and Ohio. A number of passengers disembarked who were bound for the White Sulphur and other springs across the mountains to the westward, while some were going the other way to wine-making Charlottesville or to Richmond. To the White Sulphur and other famous Virginian mountain resorts we found this was coming to be a favorite route from both north and south, its own loveliness, the opportunity of thus seeing one or both of the two great "natural curiosities" of the Alleghanian region, Luray Caverns, and the Natural Bridge, and the exceeding wildness of the scenery along the mountain division of the

Chesapeake and Ohio (or of the Richmond and Alleghany for those who choose to go *via* Loch Laird and Clifton Forge), recommending it above other routes. The Madame was very anxious to go over to the White Sulphur, which her imagination, stimulated by traditions of the antebellum aristocracy, had painted in very glowing colors ; but I told her it was impossible now, and so we kept our seats and went rushing southward again through the green hills that divide the headwaters of the Shenandoah from the tributaries of the James.

The Chesapeake and Ohio Railway, to which I have referred (or at least this part of it), was known before the war as the Virginia Central; and as it was one of the two routes between Richmond and the Valley of Virginia, it was of great importance to the Confederates. To destroy it, therefore, became one of the objects of every Federal force in the valley, though that end was not achieved until Sheridan's successes of 1864.

Toward the close of that campaign the vicinity of Waynesboro became a continual skirmish-ground, and everything was laid waste. Before the winter of 1864-5 had passed, Sheridan again appeared in force, the cavalry sent to contest his advance proving inefficient. The Confederate commander, Jubal Early, had collected his army as well as he could and posted them upon a ridge just on the further (western) edge of Waynesboro, where Sheridan's advance came up with him on March 2d. "Custer at once sent three regiments around the enemy's left flank, while at the same time charging in front with the other two brigades. The position was carried in an instant, with little, if any loss on either side, and almost the entire force captured, all Early's wagons and subsistence, tents, ammunition, seventeen flags, eleven guns (including five found in the town) and, first and last, about 1,600 officers and men. . . . As for Early, Long, Wharton, and the other Confederate Generals, they fled into the woods, and Early himself soon after barely escaped capture by Sheridan's cavalry, while making his way to Richmond. The victory at Waynesboro left Sheridan complete master of the valley."

IX.

CRAB-TREE FALLS AND THE NATURAL BRIDGE.

A Rougher Landscape.—Sources of the Shenandoah.—Crab-tree Falls.—Ascent of Three Thousand Feet of Cataracts.—View from Pinnacle Mountain.—Lexington and Loch Laird.—Approaching the Natural Bridge.—The Hotel.—Prue's Surprise.—Majesty of the Bridge.—The Attractions along Cedar Creek.—The Picture from Above.—Surrounding Scenery and Amusements.—The Bridge by Moonlight.

THOUGH the vicinity of Waynesboro, for some miles southward, is a well cultivated farming and grazing region, by the time Stuart's Draft is reached the face of the country where the track passes has become too rough for farming, and the scene from the car-windows is an ever varying panorama of rugged hills and deep ravines. Almost the only signs of human occupation are small log cabins, whose restraint-hating, indolence-loving occupants earn a scanty living by chopping logs; gathering oak and hemlock bark (one of the leading products of this region, where large tanneries exist), and sumac leaves; in hunting, fishing and feeble farming. The hills we are passing across—a tangled series of folds belonging to the Blue Ridge—are called the Big Levees, and are dominated eastwardly by the Humpback mountains. Their drainage forms the South river, and hence the uppermost source of the Shenandoah. The streams which go to make it up are countless, prattling down every green hollow. Now and then a 'pretty cascade is seen, like the Cypress falls opposite Riverside, leaping fierce and white out of the wooded precipice into a deep and quiet pool.

The greatest of all cataracts in the Virginian mountains, however, is the Crab-tree falls, reached by the old pike road from Vesuvius to Montebello and the Tye River valley east of the Blue Ridge. Sheridan once passed a large part of his army across the mountains by this road. At the very summit, from among the topmost crags of Pinnacle peak, one of the highest in Virginia, comes the Crab-tree to take its fearful course. Thence it descends three thousand feet in making a horizontal distance of two thousand, forming "a series of cascades athwart the face of the rock, over which the water shimmers in waves of beauty, like veils of lace trailed over glistening steel." The course of the stream is distinctly visible from a long distance down the face of the great crag, which contrasts sharply with the leafy masses on each side, and forms a striking landmark. The cascades vary from over five hundred feet in the highest to fifty or sixty in the lowest, and are greatly different in form and appearance. The Crab-tree is not a large stream; in one or two places the entire body of water is compressed into a shooting jet not more than six inches in diameter, but, with the economy of nature, nothing is lost in artistic effect.

Three miles down Tye river the ascent of the falls is begun by entering the forest and a chaos of massive rocks. "The forest is so dense," says H. L. Bridgman, of New York, "that scarcely can the sunlight pierce it. Stately oaks, wide-spreading maples and hickories, the birch and beech, with an occasional pine, and at rare intervals the light gray foliage of the cucumber-tree, make up a forest scene of wonderful beauty. Scarcely are we within the woods when, looking aloft, we see through the leafy green of tree tops the white spray of the 'Galvin' cataract, named in compliment to our guide, and 150 feet high. This is a clear, bold fall, and rather larger in volume and force than any of the others. The effects of the sunlight and shadow upon the fall and the forest are exceedingly graceful and picturesque, and from the beginning of the ascent all the way to the top the scene changes and shifts like a fairy panorama. . . . An hour or more of hard work and steady climbing brings us to the base of the 'Grand Cataract,' the first leap of the entire series, a clear fall of over 500 feet. It was the Grand Cataract which we had seen from the road far below, and looking upward from its base, the sight was like a sheet of foam falling out of a clear sky. The water, pure as crystal, is not projected with sufficient force to send it clear of the rock, and so it falls over its face, vailing the rugged front of the mountain as with a fleece. Standing at its base and looking upward, the spectator does not realize its immense height, but comparison of the lofty trees which tower into the heavens without approaching half the height of the falls demonstrates the fact. At the very top and crown of the fall, the configuration of the rock gives the current a sharp diagonal set which adds much to its picturesque beauty. Midway, a ledge of a few feet wide arrests the fall and throws it boldly forward in a straight line again adown a sheer and glistening precipice of more than 200 feet. At the base of the Grand Cataract daisies bloom, and the waters are quite shallow."

It is possible to work one's way upward along these capricious cataracts to the very summit, and thence overlook a wide area of primitive mountain country. All about the observers tower peaks of the first rank, heaving against the blue of heaven a surging mass of foliage. "Dotting the mountain sides in every direction are cleared fields in which corn, wheat and tobacco are raised, the clearings sometimes extending to the very summits, while scattered here and there in all directions, nestling in the intervals and pockets of the ranges, are the log cabins of the mountaineers. Safe in these fortresses and upon a kindly and generous soil, with a genial and salubrious climate, the natives live from one generation to another an easy, thriftless and contented life. No one who sees the view from the head of the Crabtree falls or Pinnacle mountain, no matter what his travels or experience in this or any other country have been or may be, will ever be able to forget its matchless charm, repose and serenity."

Through such a region as that we are now running, by the help of a thousand curves, deep cuttings or lofty bridges. Now and then wonderful landscapes open out—far views southward and westward into the richly blue folds of the mountains, but chiefly our eyes are held by green dells, the romantic river, and the captivating bits of ruined canal, which arrange themselves for an instant close to the track only to dissolve into new pictures with kaleidoscopic speed.

At Loch Laird we encounter the Richmond and Alleghany Railway, which forms an exceedingly picturesque route from Richmond westward to a junction with the Chesapeake and Ohio at Clifton Forge. Its only availability to us here would be as the means of access to Lexington, a town which southern people are fond of calling the "Athens of Virginia," because of its intellectual society and regard for books. This arises from the fact that since its foundation it has been a school town, and has now the celebrated Military Institute of which the most distinguished son was "Stonewall" Jackson, who is buried there.

"What river is this?" asks Prue after we had been tracing the pretty stream for a few miles, having passed over the divide and now were beginning to follow descending instead of ascending currents.

"The South river," I reply.

"But I thought we had just left South river behind."

"So we did. This is another, and a branch of the James. You might find a hundred South 'rivers,' 'forks,' 'branches,' and so on in the state. They were carelessly named by people who never went—"

"Natural Bridge!" shouts the brakeman, and we hurriedly gather up our baggage and alight, with perhaps the most pleasurable anticipations of the whole trip.

It is two miles from the railway back into the broken hill country, where the Natural Bridge spans one of the mountain streams. Hacks from the hotel awaited the train, and our party had soon begun the drive. A short distance brought us out upon a sort of ledge, where, some hundreds of feet directly beneath us, we could see the noble James, deep, wide and glossy, forcing its way along in the dignity of fullness and strength. On the other side a great hill rose from the water, and as we attained higher and higher levels, other ridge-like summits appeared behind, each more savage and lonely than the preceding.

The road is good and winds prettily among the hills, between a gulf on one side and tangled brush slopes on the other. It was with pleasing suddenness, too, that we emerged at last upon the broad lawns and parks of the hotel property, with its array of handsome dormitories, and its groups of smaller pleasure buildings, summer-houses and gardens. It was supper-time, and we were content for that night to sit on the veranda, listen to the ballroom music, breathe the cool, balsamic air, and sleep the sleep of weariness.

Breakfast was no sooner despatched next morning, however, than we hastened to satisfy our curiosity as to this great bridge "not built with hands," which justly ranks among America's "seven wonders."

The lawns are cleared around the head of a shallow ravine, the extreme upper point of which is occupied by an enormous mineral

THE NATURAL BRIDGE.

spring and fish basin. Down the ravine from the spring goes a well-graded pathway, which quickly disappears in the woods standing along the tumbling cascades of a brook that traverses the estate, and we follow it gleefully until it has descended three or four hundred feet into the leafy screen and rocky seclusion of one of Appalachia's most lovely glens. Prue has been sauntering on ahead, and turns a corner. As she does so we see her lift her head, a wide-eyed glow of surprise illumines her fair face, and she utters a little exclamation of delight. A step forward and we stand by her side and share her excitement—the bridge is before us!

The first impression is the lasting one—its majesty! It stands alone. There is nothing to distract the eye. The first point of view is at sufficient distance, and somewhat above the level of the foundation. Solid walls of rock and curtaining foliage guide the vision straight to the narrows where the arch springs colossal from side to side. Whatever questions may arise as to its origin, there is nothing hidden or mysterious in its appearance. The material of the walls is the material of the bridge. Its piers are braced against the mountains, its enormous keystone bears down with a weight which holds all the rest immovable, yet which does not *look* ponderous. Every part is exposed to our view at a glance, and all parts are so proportionate to one another and to their surroundings,—so simple and comparable to the human structures with which we are familiar, that the effect upon our minds is not to stun, but to satisfy completely our sense of the beauty of curve and upright, grace and strength drawn upon a magnificent scale. "It is so massive," exclaims Mr. Charles Dudley Warner, "so high, so shapely, the abutments rise so solidly and spring into the noble arch with such grace and power ! . . . Through the arch is the blue sky; over the top is the blue sky ; great trees try in vain to reach up to it, bushes and vines drape and soften its outlines, but do not conceal its rugged massiveness. It is still in the ravine, save for the gentle flow of the stream, and the bridge seems as much an emblem of silence and eternity as the Pyramids."

Descending further the path cut along the base of the cliffs, which, as one writer has said, arise " with the decision of a wall, but without its uniformity—massive, broken, beautiful, and supplying a most admirable foreground." We advance under the arch, and gaze straight up at its under side which is from sixty to ninety feet wide. It is almost two hundred feet above the stony bed of Cedar creek, but Baily doesn't remember this, and fancies he can hurl a pebble to the ceiling. Vain youth! Even gentle Prue laughs at him, and the swallows weaving their airy flight in and out from sunlight to shadow, fearlessly swoop lower and twitter more loudly, deriding his foolish ambition.

Crossing the gay torrent on a foot-bridge, we wandered up the creek a mile or more, past Hemlock island ; past the cave where saltpetre was procured for making powder, in 1812, and again during the Confederate struggle, and even penetrated the low portal within which a "lost" river murmurs and echoes to our ears its unseen history, as it plunges through

THE SALTPETRE CAVE ON CEDAR CREEK.

the dark recesses of its subterranean course ; and the farther we went the more rugged, thickly wooded and charmingly untamed was the gulch. Finally the walls closed in altogether, but finding a boat we crossed to a stairway of stone leading to Lace Water falls, where the stream leaps a hundred feet, falling in a dazzling *deshabille* of rainbow-tinted bubbles and spray.

The Bridge seen from this (the upper) side is imposing, and its magnitude is perhaps more striking ; but on the whole it is not so effective, regarded as an object by itself, as when studied from below. Harriet Martineau, who once visited the spot, and has written enthusiastically of it in the second volume of her "Retrospect of Western Travel" (1838), declares that she found most pleasure in looking at the Bridge from the path just before reaching its base. "The irregular arch," she writes, "is exquisitely tinted with every shade of gray and brown ; while trees encroach from the sides and overhang from the top, between which and the arch there is an additional depth of fifty-six feet. It was now early in July ; the trees were in their brightest and thickest foilage ; and the tall beeches under the arch contrasted their verdure with the gray rock, and received the gilding of the sunshine as it slanted into the ravine, glittering in the drips from the arch, and in the splashing and tumbling waters of Cedar creek, which ran by our feet."

Nevertheless, if you are willing to regard the great arch only as a part of the *ensemble*, and to take into just account what is around and beyond it as a proper part of the scene, I advise you to place yourself a hundred yards *above* and then observe what a charming picture of glistening torrent, flower-hung rocks, stately trees and far away mountain crests is framed into its oval ; and how incomparable is the colossal frame itself—what sublimity of design—what wealth of decoration and lavishness of color !

It cannot be too strongly insisted upon, however, that while this curious product of water erosion (slowly turning a cave into a long tunnel and then, by the falling of the most of the roof, leaving only an arch-like segment of the tunnel in the shape of a bridge) is the central attraction, there are a thousand other sources of enjoyment and pastime at this pilgrimage-point.

For those who are content with rest and gossip, fresh air by day and dancing at night, the fine new hotel offers every inducement for a prolonged stay. To the larger class which seeks more active pleasure during the summer vacation, a wide range of good roads and interesting country is open for exploration. "The Bridge," says the admirable little guide-book issued by the hotel people, "connects two of five round-topped mountains that rise boldly from the great Valley of Virginia, near the confluence of James and North rivers. These have been named Lebanon, Mars hill, Mount Jefferson, Lincoln heights, and Cave mountain, and embraced in the park. Private carriage-roads, nearly ten miles long, lead around or over them, and give through arches cut in the forest, or from open spaces, a wonderful variety and extent of mountain scenery.

"Eight hundred feet below the summit of Mt. Jefferson lie the green valleys of the rivers. Eight miles to the east the Blue Ridge, forest-covered and mist-crowned, rises to its greatest height, 4,300 feet above the sea, and extends to north and south nearly one hundred miles before it is lost in the dim distance. A little to the left the glint of broken granite alone shows where the river bursts through, and at the right the crest lowers so that the Peaks of Otter may overlook. At the south, Purgatory mountain, and at the north, House mountain, throw their immense masses half across the plain. Against the western sky North mountain, the 'Endless mountain' of the Indians, lies cold and colorless. In the lifted central space of this great amphitheatre the park is located."

An old turnpike crosses upon the Bridge, but amid the apparently unbroken forest, few persons would discover it till told by the driver. In one of his inimitable articles in *Harper's Magazine*, before the war, Porte Crayon gives a ludicrous account of how his party behaved on the brink of the chasm: and Miss Martineau confesses how her search was baffled. "While the stage rolled and jolted," she writes, "along the extremely bad road, Mr. L. and I went prying about the whole area of the wood, poking our horses' noses into every thicket and between any two pieces of rock, that we might be sure not to miss our object, the driver smiling after us whenever he could spare attention from his own not very easy task of getting his charge along. With all my attention I could see no precipice, and was concluding to follow the road without more vagaries, when Mr. L., who was a little in advance, waved his whip as he stood beside his horse, and said, "Here is the Bridge!" I then perceived that we were nearly over it, the piled rocks on either hand forming a barrier which prevents a careless eye from perceiving the ravine which it spans. I turned to the side of the road, and rose in my stirrup to look over; but I found it would not do. . . . The only way was to go down and look up; though where the bottom could be was past my imagining, the view from the top seeming to be of foliage below—foliage forever."

The driveways do not cease at the Bridge, but continue by an elevated course which gives some remarkable outlooks, and takes in various notable points.

The hotel is open all winter, and there are few days in this southern latitude when it would not be entirely comfortable to visit all the points I have mentioned, and see the Bridge under a grimmer aspect, truly, than when mantled in the garlands of summer, yet with none of its grandeur diminished.

"Well," remarked Prue, when I had read over to her what I have written, "I *do* think you have made about as great a *failure* as I have ever seen. Why you havn't BEGUN to tell of *half* the good times we had at that *perfectly* LOVELY place!"

THE ARBOR-VITÆ TREES, AND GIANTS' STAIRWAY.

"I know it," I confess with humility.

"Well, *at least*," she went on, crushing my poor effort, "I would describe the gorge seen by *moonlight*. Don't you remember, Theo, that evening when we left the hop, stole away from the crowd on the piazzas and ran down the dewy lawn together?"

"You looked like a fairy that night, Prue, in your floating lace."

"And then how we crept by those big ogreish arbor-vitæ trees, and how you laughed at me because I was a little timid in that dreadfully dark shadow under the crag; and how we tried to hear words in the tinkle of rivulets down the ledges? Then, don't you remember with what a startle of delight we came in sight of the ravine, and you said the Bridge must have been carved out of silver and ebony? Can't you tell about that?"

"No, Prue—and I shouldn't like to try. Let those who come after us find it out for themselves as one of a hundred novel joys which await the sojourner at the Natural Bridge."

X.

THE NEW CITY OF ROANOKE.

On the Bank of the James.—The Gap.—Buchanan's Iron Works.—A Town Saved by its Captors.—Crossing to the Valley of the Roanoke.—Baily's Triumphant Quotation.—Beginning and *raison d'etre* of Roanoke.—History of the Consolidated Railways.—Amenities of Roanoke.—Machine Works.—Iron Furnaces.—Stock-yards.—Minor Factories.—The Great Hotel.—Sunset Pictures.

ROLLING slowly across the lofty iron bridge which carries the track over the James at the Natural Bridge station, we skirt the base of the mountains on the southern bank, and follow closely all the windings of the stream. Not only is it impossible for the railway to leave its margin, for the most part, but through long distances it has been needful to dig into the foot of the precipitous hillside in order to make room for the tracks. On the opposite side run the tracks of the Richmond and Alleghany Railroad, following the line of the disused canal, whose broken dams still ruffle the current, and whose ruined locks are sinking into shapeless decay.

As we approach Buchanan, the hills grow even steeper, and crowd upon the river so closely that its current is greatly deepened and confined, and rushes with noisy turbulence along a lane of gigantic sycamores, willows and other water-loving trees, toward the gap where the James bursts its way through the lofty cross-range of Purgatory mountain. This gap is one which will especially interest not only the scenery hunter but the geologist; for in the northern wall of the gorge, where the river has exposed a vertical face of rock of great height and breadth, it is easy to see how the rocks there have been bent upward into an arch as high as the hill, the concentric strata in which can be counted almost

JAMES RIVER GORGE.

at a glance. Every exposed cliff and railway-cutting gives evidence to
the observant eye of how the substance of these confused knolls and
ridges has been contorted; but it is rare that so plain a cross-section of
folding is offered as in this exceedingly picturesque gap.

Between Waynesboro and Buchanan, the town which lies just above
the gap, many incidents of historical interest might have been enumerated, and the names mentioned of many great men who were its sons;

NEAR BUCHANAN.

but no consequential operations of either army in the late war occurred
there. At the latter town however, began a series of very memorable
scenes.

On the evening of the 14th of June, 1864, Buchanan was noisy with
furnaces, forges, foundries and mills, especially the powerful branch of
Tradegar Iron Works, where cannon, ammunition, and other iron-sup-
plies were cast for the Confederate government. Here were flouring
and blanket mills also, and in the neighborhood lay farms producing food
and forage for the army. In the town, as guard, was McCausland with
the cavalry which had just come back from disasters before Sheridan.
Demoralized and weak, these troopers were dismayed to hear that the
Yankees were just across the river in great force, and would capture
them all in a hurry. The river was easily fordable here, but McCausland (the same who set fire to Chambersburg and several Maryland villages), saw fit to burn the bridge against the protest of the citizens.
From the burning bridge houses caught fire, and the whole town would
have been destroyed had not the Yankee soldiers turned firemen and
helped extinguish the flames. This salvage accomplished, the captors
(Hunter's fifteen thousand raiders) destroyed the ordnance factories

which were so valuable to the Confederacy, and pushed toward the Peaks of Otter, " at a great expense of pioneer labor and bush-fighting."

The James river, at Buchanan, passes close to its southern watershed; and having crossed the ridges which closely beset the town in that direction we are free from the grasp of the sterile and jungle-covered hills and descend into the valley of the Roanoke, through the farming and fruit raising districts of Houston (the boy-home of Sam Houston, of Texas

CROZIER IRON WORKS.

fame), Troutville and Cloverdale. Seventy thousand apple trees were planted in Cloverdale alone during 1883; and—

"Cut it short!" Baily calls out with that disrespect for his elders which will be the death of him some day. "Here's our guide-book telling us all about it. Listen to this:

"We enter the Roanoke valley amid scenes of surpassing beauty. The setting sun purples the tops of the mountains and throws its slanting rays over the rich field and pasture lands; the twilight steals out of the forest and dims the blue thread of mist along the James; the cattle low in the shaded lanes, the sheep-bells tinkle on the hills; Æolian winds ring among the dusky trees,

' Night draws her mantle and pins it with a star!'

"The city of Roanoke blazes up ahead like an illumination; red-mouthed furnace-chimneys lift like giant torches above the plain; the roar of machinery, the whistle of engines, the ceaseless hum of labor and of

A MOUNTAIN RIFT NEAR ROANOKE.

life in the very heart of a quiet, mountain-locked valley! We roll into the finest depot in the state, and are escorted to a hotel that would do credit to the proudest city. We tourists go to bed dumbfounded!"

"That's the way to do it!" cries Baily, closing his book in triumph.

And that's just the way we did.

The nucleus of this city of Roanoke was a small village known as the "Lick," where a salt lick, or saline impregnation of a piece of marshy land, originally attracted the wild animals of the vicinity, and, with the advance of settlement, the domestic animals of the pioneers. It was on a post-road, and had a tavern, store and post-office, but is now simply a suburb tenanted wholly by negroes. The country round about was exceptionally rich in agricultural land and forest growth, and soon attracted settlement and cultivation. On the opening of the Virginia and Tennessee Railway, in November, 1852, the business of the neighborhood naturally gravitated to the immediate vicinity of the line, and a town was started about the railway station called "Big Lick," half a mile distant from "Old Lick," which finally became a hamlet of about 600 people.

In 1870, the Virginia and Tennessee, by consolidation with its connecting lines, became the Atlantic, Mississippi and Ohio Railroad, and this having become embarrassed in its finances was purchased by a syndicate of capitalists in Philadelphia, most of whom were already interested in the Shenandoah Valley Railroad, then in course of construction. It was decided to continue the latter line to a junction with the former at Big Lick (achieved in June, 1882), and operate them in association. The name of the Atlantic, Mississippi and Ohio Railroad was changed to Norfolk and Western. An operating arrangement for twenty-five years was concluded in September, 1881 with the East Tennessee, Virginia and Georgia Railroad, and its leased lines, and the Shenandoah Valley Railroad, and the entire system of 2,203 miles of railway has since that date been worked in entire harmony in all matters of general traffic, as the Virginia, Tennessee and Georgia Air Line. Economy and efficiency necessitated some central point for the control of the Norfolk and Western and Shenandoah Valley Railroads, the head-quarters of their direction, position of the shops for construction and repair of equipment; and residence of many of their employees. A company was therefore formed, which gradually bought several thousand acres of land around the junction, nearly all of which was farm land, procured the legal authority and laid out a town site, which was named Roanoke after the river which flows half a mile southward.

This was in the fall of 1881. Now Roanoke is a town of lively business appearance, and of new, modern, and in many cases very handsome houses, with a population of seven or eight thousand and more coming. Its streets are lighted by gas, and the whole town is supplied with sweet, pure water drawn from "Big Spring" a mile and a half away, which is

OFFICES OF THE CONSOLIDATED RAILWAYS AT ROANOKE.

one of the most picturesque spots in the valley of the Roanoke river, whose lively current purls near by. The town contains a number of churches, good schools, a library association, an opera house and various other means of mental and moral culture, as well as of material progress ; while the presence of so many executive officers and their families, presupposes a society of more intelligence and social experience than is usually observed in so new a town.

"The requirements of such a population," says a recent report shown me by the indefatigable Baily, "almost entirely consumers, and the position of the city, at such an important railway junction, surrounded by an agricultural territory of such great productiveness, with abundance of iron ores on every side, vast supplies of coal and coke within easy distance, and such a nucleus of manufacturing industry already established, seem to confirm the promise of a prosperity built upon the most solid foundation, and capable of indefinite expansion."

The largest element in the progress of Roanoke was the building of the Roanoke Machine Works, which owns a large tract of land and has constructed extensive buildings in the angle between the two roads. These buildings consist of brick shops, engine houses, and mills,

BIG SPRING, NEAR ROANOKE.

locomotives, stationary engines, cars of every grade and description, covering many thousands of square feet, and supplied with all the ponderous and complicated machinery necessary to make all sorts of bridges, and all kinds of cast or forged iron work. This does not mean merely that the machinery or cars may be put together here ; but, except a few specialties, every part of the locomotive or car, from the wheels to the last ornament, is made and fitted as well as "set up" here. It would be out of place in a pamphlet of this nature to give an extended description of such works, to which these railways look for nearly all their rolling stock ; but the visitor to Roanoke will find it well worth his while to go through them.

The raw material of iron and steel used is largely supplied by the Crozier Steel and Iron Company, whose blast-furnace is a quarter of a mile away, and another object of interest to tourists, who often go at night to witness the thrilling spectacle of drawing the molten iron from

HOTEL ROANOKE.

the furnace into the molds where it will be cast in "pigs." This company derives its ores (brown hematite) mainly from the upland mines owned by it near Blue Ridge station, ten miles eastward, and from the Houston mines, fifteen miles northward. The yield sometimes reaches a hundred tons a day, and the greater part is marketed in Pennsylvania in successful competition with local manufacturers. Another similar enterprise is the Rohrer Iron Company, which owns extensive deposits of high grade limonite ore half a dozen miles south of town. This property is reached by a narrow gauge railway, which may ultimately be extended through to the Danville and New River Railroad, in North Carolina, and at its terminus in West Roanoke the company owns land upon which it now stores and ships its products, and will probably construct a furnace. Near there are the Roanoke stock-yards, where abundant conveniences for the transferrence of cattle are provided, together with a hotel for the drovers and traders having telegraphic communication with northern markets.

In addition to these large concerns many smaller ones contribute to the prosperity of the place; such as tobacco factories, lumber-working mills, cigar-making shops, spoke factories, bottling works, and the like. So rapid and persistent has been the growth of the little city, the site of which three years ago was all a wheatfield, that although the Town Company has expended $600,000, its profits have been very satisfactory.

For the equestrian, the vicinity of Roanoke is full of opportunities. A hard, even road leads away eastward over the ridge, where most of the handsome homes of the residents are built, and brings us to the Big

TINKER AND MILL MOUNTAINS, ROANOKE.

Spring, a fountain-head of water sufficiently powerful to run the huge wheel of a flour mill, and to supply the city with a plentitude of the purest water.

To the westward other roads wind away into the hills. Under the pilotage of two genial citizens we made a saddle journey of discovery in this direction. We found, hidden away in the peaceful seclu-

1.—A SHADY PORCH.   HOTEL ROANOKE.   2.—MAIN STAIRWAY.

sion of a pretty valley, the Hollins Institute, a popular seminary for young ladies, always filled with merry and bright-eyed maidens from every state of the South, under tuition of an excellent corps of instructors. Two miles beyond we came upon one of those sermons in stones which are as an open page to the geologist—a rift in the ledge where a little fretful stream poured down between the rocky jaws over the ruins of a log dam and past the remnants of a flume and mill—as pretty a bit of rockscape as one will find in these mountains. Here the pent-up waters of a vast inland lake have some time burst through and scattered the fragments of the massive gateway right and left through the valley. We found just time to make a hasty sketch, and retraced our steps to the Institute, beneath the hospitable roof of which we tarried that night.

Returning to Roanoke in the morning by the mountain road, our artist halted to add the bold outlines of Tinker and Mill mountains to his sketch-book, and we wished, when we drew rein at the hotel an hour later, that our ride had been twice as long.

The three buildings which catch the eye of the traveler, and surprise him, are the railway station and its "low-ceiled, dainty" eating-house in the Queen Anne style—though, as Charles Dudley Warner

LOBBY OF THE HOTEL ROANOKE.

said of it, that queen probably never sat in so taseful a dining-room or had so good a dinner; the railway head-quarters, falling in a cataract of peaked roofs and balconied fronts down the slope of the street; and the splendid hotel crowning the hill in the midst of lawns, parterres of flowers and ceaseless fountains. In the presence of the accompanying illustrations it would be superfluous to describe their outward appearance.

Interiorly—to speak now of the Hotel Roanoke,—the wood-work is hard pine, finished in the natural grain; the furniture ash and cherry, and all the arrangements tasteful as well as commodious. The parlor is as pretty a room as you will find in many a mile, and the dining-room light and cheerful, with small tables and growing plants. Under the same management as the Luray Inn and a leading hotel in Philadelphia, the table and service are of a high order; and I do not know a better resting place for the tourist than this. All this may seem high praise for a hotel, but it is given ungrudgingly. We spent a good many pleasant days there and paid for them squarely; hence I can say what I please, and sum it up in the candid opinion that Hotel Roanoke has nothing to approach it (save at Luray) between Philadelphia and Florida.

There was a certain corner of one of the upper piazzas a little out of the way, where we used to like to sit an hour or so after tea, smoking our evening cigars, watching the glories of the sunset, and discussing things in a hopeful strain that would have vexed Michiavelli to the soul. The mountains stand in an irregular circle about Roanoke, none too near for the best effect, and the western view is an especially fine one. The lowering orb of light sinks grandly behind the line of mountain wall, across whose serrations its last rays gush in a blinding effulgence which slowly pales away through every rosy and nacreous tint into the sweet twilight of the summer night. I remember a remark by Prue, that the day here was like the fabled dolphin which in its death put on a shimmering robe of swiftly changing colors, and so passed away gloriously. Nor is the beauty all in the sky, for the foreground is, nearest, the picturesque structures of the town, then a billowy stretch of green and bosky knolls, and finally the obliquely retreating array of the Alleghanies, where

"headland after headland flame
Far into the rich heart of the West."

Sitting thus one evening, I am asked:

"Were you ever at Norfolk?"

"Oh, yes." I reply, "and had a capital week of it, too."

"Did you pass over the Norfolk and Western between here and there?"

"Pretty nearly; I came as far as Lynchburg."

"Tell us about it," is Prue's request. "I feel a great deal of interest in Norfolk on account of its strawberries."

"And I on account of its peanuts," mimics Baily, at which the young woman near him makes a little *moue*."

Railroad Bridge

XI.

NORFOLK AND PETERSBURG.

Virginia's Great Seaport.—Commercial Advantages.—Colonial History.—Revival after the War.—Cotton.—Peanuts.—Garden Truck.—Oysters.—General Supplies.—Lambert's Point Coal Wharves.—Old Point Comfort and the Harbor.—Hampton.—Ocean View.—Virginia Beach.—The Dismal Swamp.—Baily's Impertinence.—Petersburg.—Forts Hell and Damnation.—Peach-trees as Monuments of Battle.

"NORFOLK," I say, in response to the question which closed the last chapter, "is the most business-like city in Virginia, and next to the largest. It has a right to be so, because of its situation. It is just at the mouth of James river and the Chesapeake bay, and has a large, deep and well protected harbor, where any kind of shipping can enter without delay or danger. The government has had a navy yard at Portsmouth (which is an

THE MARKET SQUARE AT NORFOLK.

attachment of Norfolk) for many years; and Hampton roads, just below the city, is the favorite ground for naval reviews, etc. She has lines of ocean steamers to Liverpool, and to Boston, Providence and New York. Two or three lines of steamboats connect her with Baltimore, of which the celebrated 'Bay Line' is the best known; and steamers run regularly to Washington, Richmond, and up all the lesser rivers, as well as southward through the canals. She is the terminus of a railway to the southern coast-region, and of this great east-and-west highway—the Norfolk and Western. She has every commercial advantage she could wish, therefore, and a clientele of about fifty thousand people."

"Isn't it an old, old town?" Madame Prue inquires.

"Very old. It was almost the first spot upon which colonists set foot, and has been recognized as a settlement for two centuries or

NEW COTTON COMPRESSOR AT NORFOLK.

more. In Norfolk, to-day, you may find some of the quaintest and most typical homes of the old fashion which remain anywhere in America; and I do not know a seaport on our coast more picturesque along its water front. It is a charming place for a stranger to stroll about in, and, when he becomes acquainted with them, he finds the people warm-hearted, intelligent and delightful. There was surely a pleasant omen in the fact that the first settlers of prominence were named Wise and Thorogood!"

"Norfolk was an important point during the rebellion, if I remember aright."

"It was the great naval centre. The city was not destroyed, only paralyzed, by the war, and after its close the citizens returned and began to pick up again the threads they had dropped. They found it needful, however, in the new order of things, to make many innovations.

OLD CHURCH AT NORFOLK.

Among the first attempted was dealing in cotton. Started by Mayor William Lamb's ventures in 1865, it advanced until, in 1874, the railways began giving through bills of lading via Norfolk to foreign destinations. Then followed arrangements for the proper handling of cotton at this port. Steam compressors of great power were built, a cotton exchange was organized, with every facility for business parallel with the New York exchange, and a great many men gained their livelihood by trading in or handling this staple. During the season of 1882-3 nealy 800,000 bales came to Norfolk, about half of which was consigned through on foreign bills of lading. Norfolk is now the third cotton port in the United States in point of receipts, and second only to New Orleans in point of exports to Great Britain. This result has been possible by the concentration there of lines of railway transportation for receiving, and of ocean steamers for distributing. Among the former, the Norfolk and Western brings three-fifths of the total receipts, gathered from as far inland as Memphis and Atlanta. Now, who was it expressed an interest in peanuts?"

"Madame Prue," says Baily, with unblushing effrontery.

"Well, peanuts, next to cotton, make the largest item of Norfolk's trade. They have been grown since the war in all the tidewater counties of Virginia, and somewhat also in North Carolina and Tennessee. The farmers choose a light soil, manure with marl, plant in May, 'cultivate' the rows of plants assiduously during the summer, and harvest in October. The vines, after being thrown out of the ground, are stacked in the field and left for from ten to twenty days, when both vine and nuts will be cured. The nuts are picked and sent to market; the vines are saved to be fed to cattle. The peanuts are sold to buyers for factories in Norfolk—there are factories also in Petersburg and several other places —where they are sorted into commercial grades of quality, and put through machinery which cleans them thoroughly of all earth, and polishes the shells into a fit condition to be roasted. These processes are very interesting, and anyone visiting the city ought to try to see the operation. During the past season the crop amounted to a million-and-a-half

TERMINAL WHARVES AT LAMBERT'S POINT.

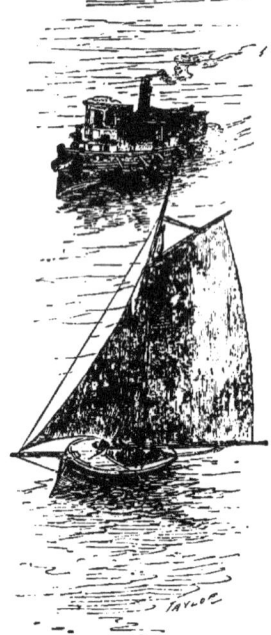

of bushels; and the yield of the present year (1884) is expected to be two-and-a-quarter millions, two-thirds of which is produced in the Old Dominion."

"Now tell Mr. Baily something about his friends, the cabbages," remarks my wife, sweetly.

"Meaning garden truck, generally, I suppose. He'd better read Mr. E. P. Roe's book about small fruits, which deals principally with that district and covers the case. In the spring, enormous quantities of vegetables and fresh fruit are sent from here to northern cities, as everybody knows who does any shopping in New York or New England—eh, Prue?"

"Certainly, and we get no end of oysters from there, too."

"You bet!" Baily exclaims. "Who doesn't know Norfolk oysters—especially the rich, rare and racy Lynnhaven Bays!"

"Well, those are the chief 'points' on Norfolk as a business city.

How she becomes a supply depot for a wide extent of country you can easily understand from what I have said of her position and traffic facilities. She has just had completed the construction of splendid piers,

IN FORTRESS MONROE.

elevators, coal-shutes and other terminal facilities at Lambert's Point, four miles from town, whereby the Norfolk and Western Railroad can receive and transmit freight from ships, not only at convenience and expense far more advantageous than at present, but can handle the unlimited quantities of coal which are now being mined in the mountains and carried to the seaboard for use by steamers there, and for export to distant markets."

"Any chance to have a good time in Norfolk?" Baily asks, biting off the end of a big cigar, but neglecting to give me one until I remind him of the impropriety of his conduct.

"Good time? Why, of course. Besides all the opportunities for pleasure 'within its gates,' that belong to a lively southern city, there are the peculiar local opportunities which its proximity to the ocean and bay afford."

"Dear me!" murmurs, as if to herself, "what fine phrases!"

Prue has a small book in her hand and is not listening very attentively, but I don't mind; I am used to it.

"For instance?" Baily inquires.

OLD POINT COMFORT, FROM SOLDIERS' HOME.

"Well, just listen to a paragraph or two out of a little pamphlet that somebody has issued in regard to the attractions in that neighborhood:"

"Passing from Norfolk to the ocean, the traveler sees the Naval Hospital, and the spacious and magnificent Navy Yard at Gosport, which the Federal troops attempted to destroy when they evacuated the city in 1861. Their success was only partial, and the works have since been in active operation. In this yard the Confederate ram, 'Merrimac-Virginia,' was built, which destroyed the frigates 'Congress' and 'Constitution' in Hampton Roads, and had the famous fight with the 'Monitor.'

"Other points of interest are Old Fort Norfolk, constructed in 1812, where a magazine of supplies was afterwards established; Craney Island, where the engagements of two wars were fought; Sewell's Point, named for one of the earliest colonists; the Rip Raps and Fortress Monroe.

"Other points of interest lie still further below: there are Lynnhaven Bay, celebrated for oysters; Cobb's Island, famous for bathing; Cape Charles and Henry, and other historic places.

"The Hygeia Hotel, Old Point Comfort, Virginia, is situated at the confluence of the Chesapeake Bay and Hampton Roads. It is washed on three sides by broad sheets of salt water. The climate, in consequence, is mild and soft during the winter and spring, and in the fall vegetation continues untouched by frost long after the inland has been blighted. The heat of summer is mitigated by the constant sea breeze and the superior ventilation of the building.

"The hotel itself is new and spacious, accommodating one thousand guests and open all the year. The guests have numerous amusements— fishing, sailing, bathing, boating. One minute's walk will bring one to Fortress Monroe, the largest fort in the United States ; a quarter of an hour in another direction to the National Cemetery, the Soldiers' Home,

VIEWS FROM DOCK PAVILION, HOTEL WARWICK.

and the Hampton Normal and Agricultural School for Indians and colored people."

"There's at least one important omission in that list," cries the watchful and well-informed Baily.

"What is that?"

"Newport News and the Hotel Warwick, where the Old Dominion steamers from New York land. The writer speaks of the fight between the 'Monitor' and the 'Merrimac' as a feature of interest at Old Point Comfort; but it belongs much more to its rival, for that fight occurred right off Newport News, and a long wharf has been built out from the shore directly over the wreck of the frigate 'Cumberland,' which was sunk by the Confederate ram early in the encounter. The Warwick is a new hotel, of brick, very ornamental within, and in general

BOWLING HALL, HOTEL WARWICK.

quite as handsome as the Hygeia. It stands close to the beach, commands a fine view of the shipping in Hampton Roads, and has beautiful and historically interesting surroundings. It's a toss-up with me whether I'd choose the Hygeia or Hotel Warwick, if I were lucky enough to be ready to go to either just yet."

"Are there no seaside pleasure-places on the Norfolk side of the water?" Prue asks me.

"Oh, yes. There's Ocean View, for example, ten miles north of town, where for many a year the beauty and chivalry of Norfolk have gone out to spend a week or an hour, as chance served, on the shore of Hampton roads, fishing, and eating crabs all day, and eating crabs and dancing all night. Little open-car trains run back and forth every hour or two during the warmer two-thirds of the year, and it is no end nice after a warm day to take the girl of your heart, dash out there in the cool of the evening, meet a lot of friends in the spacious ball-room, where special dress is no object, and then skip home, one of a jolly company, by the ten o'clock train.

" But that is not on the ocean, is it, in spite of its name?"

" No, you get a glimpse of it down through the Capes, but for an interview with the great briny itself, you must go out to Virginia Beach. That is twenty miles due east of the city, and reached by a narrow-gauge railway lately built. The depot is the same as the magnificent new station of the Norfolk and Western, which is the admiration of all who enjoy fine architecture; and the ride is an entertaining run through a district having many historical associations. Then the beach itself (you come upon it very suddenly and strikingly) is as fine as you could see anywhere in the world. Miles and miles up and down the coast stretches this mall of hard sand and those flashing lines of brilliant breakers. It is a wonderful sight, because of its vastness—the whole breadth of the open, unhindered ocean in front, the mystery of the unbroken and apparently primitive forest behind, and, between, this golden and white margin of coast, straight as a ray of light and far reaching as the eye can follow!"

"Any hotel there?"

"A very fine one of huge size and pleasant equipment. It stands as near the surf as safety will allow, and has several acres of piazzas and

SCENE ON VIRGINIA BEACH.

'pavilion' attached. I believe a battalion drill could be held in that great pavilion, which is crowded during the day with excursionists from town, and haunted at night by the sensible few who wait for the late train home, or by the habitués at the hotel, who stroll in the half-light on the seaward side, watching the luminous surges. I never missed you more, Prue, or felt that you had lost so much in one evening, as when I spent those twilight hours alone at Virginia Beach."

"Now," said Baily, after a complimentary pause. "Tell us about the trip hither."

"You leave Norfolk comfortably in the morning, go through Petersburg at noon, get to Lynchburg toward sunset and come on here to Roanoke for tea."

"So the time-table tells me," was Baily's dry retort. "What is there to see along the road?"

"Well, south-eastern Virginia is flat, truly, and less entertaining than the mountain country; and if one has to travel through its pines and scrub oaks for several days in succession, as I did when I went along the coast last year, he gets extremely tired of it; but though 'flat,' the ride westward from Norfolk is by no means 'stale and unprofitable.' As soon as you come past Suffolk, a dozen miles inland, you start upon what I dare say is the longest railway tangent in the world, for there the track runs absolutely straight for fifty-six miles. You could look

HAMPTON ROADS.

unimpeded from Suffolk to Petersburg, if perspective and curvature of the earth permitted."

"How is this tangent possible?"

"The land is poor, or half submerged, and the region almost uninhabited. The first few miles is run through the northern edge of the Great Dismal Swamp, and is a monument of skillful engineering and patient work, the credit of which belongs to General and ex-Senator Mahone. That swamp, by the way, is something worth taking much trouble to see."

"How do you get at it?"

"Stop at Suffolk (unless you can get a train which will let you off at the right spot), and walk two miles down the track to the Jericho Canal crossing. There you will find some negro (get Ike "Chalk" Winslow if you can) who will take you in a cypress canoe through the long, narrow, overgrown canal as far as you like. To Lake Drummond, the great pond which fills the interior of the swamp, is ten miles, but by starting early you can go there and get back by dark. It is an extremely interesting trip through the most novel scenes ; and if your boatman is one of the old swampers, who has spent his life in cutting and boating juniper logs and cypress shingles from the recesses of the vast morass, you can draw from him many a legend and bit of slave-history or curiosity of woodland experience."

"Are they not arranging to reclaim a large part of the swamp," Prue enquires.

"Yes, but that is too long a story to go into now. The whole area of the morass (which contains about 150,000 acres) is from fifteen to twenty feet above the level of the tide, and it only requires to cut certain drains

NEW STATION OF THE NORFOLK AND WESTERN RAILWAY AT NORFOLK.

in order that the water may run off. It is held back now by the spongy peat and the tangled masses of vegetation, roots, buried logs, mosses and ferns, which cover the real bottom with a layer many feet in depth, out of which the water cannot find its way. A plan of drainage was sketched by Washington, and a company formed before the Revolution to carry it out; but the wars and other matters prevented. Now this same old corporation has been revived, and the reclamation of the northern half of the swamp will no doubt be made within a short period."

CANOEING IN THE DISMAL SWAMP.   A SUFFOLK FARM-HOUSE.

The next point to which I "called the attention of my listeners," as the preachers say, was Petersburg, where the railroad from Richmond to Weldon and the Carolina coast cities crosses the Norfolk and Western. Petersburg is the most important town in south-central Virginia and has a wide reputation through its tobacco manufacturing, for it is the centre of a tobacco growing region. The factories there are devoted chiefly to the making of "plug" chewing tobacco, and strangers can easily get a sight of the interesting processes and machinery by which it is prepared. The trade of Petersburg is almost wholly in exporting, and one great house finds in Australia and the South Pacific islands its largest market. The tobacco called for by that trade is strong, black and compact beyond anything which is liked in America, and is sent out in bond, to the disgust of the Internal Revenue Collector of the District.

Petersburg is also a centre for the spinning of cotton and woolen clothes ; for the grinding of sumac, the leaves of which reduced to a powder have come, since the war, to have a high commercial value as an agent in tanning fine kinds of leather. The gathering of these leaves, which are extremely plentiful in the mountains, affords almost the only means open to a large class of poverty-stricken backwoodsmen, for purchasing the few "store-goods" their simple ideas of life require ; while to many a hard working wife and child, in better circumstances, the sumac gives pin-money which otherwise would be lacking. Peanut mills, breweries, fertilizer factories and various other factories flourish in Petersburg, where also is done a large jobbing business with country merchants.

The chief interest in the place to the tourist, however, arises from its history, both Colonial and that made in the more recent and deadly years when Grant besieged it with the army of the Potomac, and Forts Hell and Damnation earned their fiery titles. Room does not suffice to tell

FOOTPRINTS OF WAR.

here the long story of how the Federal grip was slowly and relentlessly tightened about the fated town, and how, day after day, week in and week out, the city was under artillery fire. The tale can be read in any history, or heard from a thousand witnesses in that region. The fortifications stand in fairly good order, and are the object of every visitor's first interest. Time and the plow have leveled some, but their contours may be traced, by lines of peach trees if by no other sign,—trees planted by the soldiers as they lay in the trenches and furtively nibbled the juicy fruit. Most interesting of all is "the crater," where that great mine was exploded which was intended not only to make a breach but to destroy a garrison. The railway passes just underneath it, and along the line of Grant's front, where the terrific charge ensued which marked that farm-slope as one of the bloodiest fields of the Civil war. A ghastly museum of relics has been brought together at the crater, exhibiting with mute eloquence the awful fury of that hour.

XII.

IN THE VALLEY OF THE JAMES.

Approaching the Mountains.—War Recollections.—High Bridge.—Farmville and its Colleges.—The Tobacco Region.—Lynchburg: a City of Terraces.—Railways and Trade.—Tobacco Factories and Market.—Iron and Iron Mills.—" Sidehill Critters."—Peaks of Otter.—Hunter's March upon Liberty.— Railway Destruction as a Military Science.—Fighting in a Burning Forest.—The Attack upon Lynchburg.—Hunter's Leisurely Retreat.—Blue Ridge Springs.—The Pretty Girls of Coyner's & Gishe's.

FROM Petersburg the road rises rapidly toward the mountains, and passes a country replete with military associations.

Burkeville, the first station of consequence, and in the midst of a fertile region, is the junction of the railway between Richmond and Danville, and passengers from the West change cars here for the capital of the state. Proceeding toward Farmville, the face of the country grows less regular, the soil improves and interest grows. Every station and roadway along this part of the line has its war story to tell. Jetersville, Sailor's Creek, Fort Gregg, Five Forks, where the Confederacy made its final fight, and Cumberland Church where, in a sharp skirmish, the Federal forces suffered their last repulse. Just beyond Cumberland Church is the High Bridge, which was in olden times the terror of travelers, but is now an iron structure of the most massive character, a mile long, spanning a depression once evidently the bed of a lake, but now rich with corn and tobacco. The latter crop is the staple production of the region, which is especially suited to it. From the High Bridge a serenely beautiful landscape is spread before and beneath the eye; its horizon formed by the varied outlines of the distant and always admirable Blue Ridge. In this vale, now so sunny and peaceful, happened one of the most impulsive cavalry fights of the war, where horses dashed breast to breast, and sabre clashed against sabre, in the fury of hand to hand conflict.

Farmville, the centre of this fine agricultural region, is a community not only of trade, but of peculiarly intelligent and cultivated people, typical of the best of those rural social *ganglia* which were the pride of Virginia under the old regime. Near here stand Hampden Sidney College and the Union Theological Seminary (Presbyterian), besides a popular watering-place called the Farmville Lithia Springs.

Next comes Pamplin's Depot, where that celebrated clay pipe is made—"the great nicotine absorber, which excels all the meerschaums of the world." Not far beyond is Appomattox station, near that world-renowned court house where the army of "tattered uniforms but bright muskets" surrendered its flags to the Union. A little farther we emerge from the hills which have gradually grown around us, to move out upon the bank of the broad James river, and after following its picturesque bendings a few miles, pull up in the union station, under the walls of Lynchburg—exactly midway between Norfolk at one side of the state, and Bristol on the extreme of the other, for it is 204 miles to either boundary.

Not content with my account of this town, or better, inspired by it, Prue and Baily declared they wanted to see it for themselves, so we all made an excursion thither just before going westward from Roanoke.

Lynchburg is well worth seeing, and your pleasure is accompanied by the satisfaction arising from what is well earned. The James river passes through a group of hills at this point, which begrudge it room and rise steeply from its edge. It would be hard to find a more unsuitable place to set a town ; yet here has grown up a city of between fifteen and twenty thousand people, half of whom can look down their neighbors' chimneys. The railways and a few mills, by the help of excavations and bridges, have made room enough to lay their tracks and build a station down near the river level, but all the rest of the town clings precariously to some steep hill-side. If you walk up from the station, you climb a series of staircases ; if you ride, your omnibus is drawn by four or six horses. When you leave your hotel and walk abroad you must choose between going up hill or down, though some streets chiseled along the hillside—all the houses on one side having high porches, and on the other all the gardens dropping away from a low rear basement—run fairly level for short distances. These are the principal residence streets, and lie tier above tier as at Quebec, Duluth or Brattleboro. At intervals a cross street rises from one to the other, not too steeply for horses to use; but in many cases wooden stairways, or zig-zag paths are alone available. An ornamental improvement has recently been made in one such useless street by turning it into a park, dropping steeply from the quaint old court house to Main street.

This terraced arrangement offers many advantages to the architect and landscape gardener, and many beautiful and picturesque homes meet our eyes as we stroll about, delighted more, perhaps, with the architectural examples of ante-bellum affluence than with the brand new modern houses scattered among them. The loose soil and upright position made thorough paving necessary, if the whole place was to be kept from washing away, and this lends to Lynchburg a clean and thrifty aspect wanting in most of its rivals. From any one of the many high points, and from hundreds of pleasant windows overlooking the lower town and the river, charming views are presented. "Opposite are the bold cliffs of the James; far to the east the river loses itself in green meadows, and behind dim woodlands; out in the westward the blue hills climb skyward, and the famous Peaks of Otter prop up the feathery clouds; southward the panorama opens with a glint of glory on wooded hills and misty valleys, and shuts out the view only when the eye pauses at the dropping horizon."

Lynchburg flourishes chiefly, perhaps, because it is at the centre of the rich Piedmont tobacco growing district; and is also a depot for iron.

TOBACCO WAGONS AT LYNCHBURG.

To this has been added a concentration of transportation facilities which have caused it to become the head-quarters of wide trading with farmers and country merchants, almost without a competitor between Richmond or Norfolk, and Knoxville. Through this James River gap, long ago,

NEGRO WAGONERS. AN EBONY GABRIEL.

was built the James River and Kanawa Canal, now the road-bed of the Richmond and Alleghany Railway. Railways also connect the town with Washington, Norfolk, Danville and Knoxville. These bring hither so much tobacco (not to mention lumber, tan-bark, sumac, grain and general produce) that seventy or eighty establishments are engaged in its manufacture or manipulation, and the town has become an important purchasing point for northern factories. Many of the Lynchburg brands of tobacco (especially that made for smoking) have an old and world-wide reputation, and others are gaining newer but equal fame. As for iron, one furnace makes thirty or forty tons of iron a day, drawing its ore from the near neighborhood; and several foundries, machine shops, railway repair shops, a nail mill, and other similar enterprises are in operation.

THE PEAKS OF OTTER.

Ample water-power is afforded by the James, which here descends with rapid current, turning hundreds of industrious wheels, and sure to be called upon to turn many more in the near future.

On the whole, our short visit to Lynchburg was productive of much amusement and instruction. As Baily says: "It is a nice little city, six or seven stories high."

A railway ride, extremely pleasant at *any* time, carries one from Lynchburg to Roanoke, but Prue and Baily and I found it especially so on the brilliant, well-washed morning when we undertook it before the heat of the day had arrived.

The track clears its way through cuts and tunnels, spans lofty bridges and runs along steep declivities devoted to pasturing what must surely be "side-hill critters," with legs on one side shorter than those on the other, until it has escaped the jumble of hills that environ the three-storied town, giving many charming outlooks by the way. Soon the Blue Ridge comes plainly into view on the right, where two sharp and prominent heights, easily dominating the range, loom up ahead and catch every eye. They are the Peaks of Otter—the loftiest points (because of the resisting hardness of their syenite frames) in all the Blue Ridge. Prue has possession of a guide-book and finds that it has been ahead of us in experience:

"We sweep along," she reads, "through fair meadows, green valleys, by orchard and woodland, through fields of corn and patches of tobacco; we see the "mica flakes" in the railroad cuts, notice the red iron stain on the hills; we scare the fat cattle in the low lands, and waken up the well-to-do farmers from their *siesta* under the shade trees.'

"Do you remember," I say to my companions, "that when we were at Buchanan, which is only a dozen miles straight north of here, I described how Hunter's raiders seized the town and destroyed the mills and ordnance foundries?"

"Yes."

"Well, he marched from there the next day straight across the Blue Ridge between those two peaks, where a tier of rich grain and fruit farms fills the saddle."

The retreating Confederates had felled trees across the narrow road, blown it away where it ran along the edge of precipices, tumbled down masses of overhanging walls in the depressions, and by bushwhacking at every step obstructed the Union advance and caused great loss to their trains. Hunter pressed on, however, and having won his way through the defiles was substantially unopposed on the march through Fancy Farm to Liberty,—a jolly little tobacco making town, a dozen miles from Lynchburg, which the train was just now entering, and where a summer hotel and half a hundred summer dresses welcomed us under the trees at the station.

Here, at Liberty, the raiders struck this railway, which then, as now, ran down into East Tennessee, and which was of the highest importance to the Confederates as the great avenue and resources for supplies and for the transference of troops. To break it was a measure of strategy and its destruction a legitimate act in war. Colonel Halpine, the chronicler of this expedition, tells with what system and vigor the work of destruction was done. Up went the rails for miles and miles along the road; the ties were gathered and set on fire; the rails laid across them until heated in the middle enough to be bent out of all shape; the torch was applied to trestles and bridges of wood, while bridges of stone or iron were "sent kiting" by gunpowder.

Marching from Liberty toward Lynchburg, obliterating the railway as they advanced, the army halted for the night on the Big Otter, after a desultory encounter with the enemy at New London, where are the still famous Bedford Alum Springs. So slow was progress possible through that rough and wooded region, that at 2 P.M. the next day they had only reached Diamond hill, where the Confederates were entrenched in barricades. This was on the wagon road, a short distance south of the railway. The barricades were taken by a charge, and so badly demoralized were its pseudo-defenders, that " had it not been for the rapid coming on of night," writes a Federal staff-officer, " and the necessity of removing our own and the enemy's wounded out of the woods, which had caught fire during the action, and were now burning fiercely with a mighty crackling and roar, only pierced by the terror-stricken screams of the mangled men who lay beneath the flaming canopy of beams and branches, we might have pushed on into Lynchburg."

Next morning, however, it was too late. An attempt had been made by the Union soldiers to destroy the railroad bridge across the James, but it had failed, and all night long Lee's reinforcements were poured into the city. The forts and breastworks that crowned all the hills—you may see their remains—were strong and well manned. Hunter's legions charged but were repulsed. A battalion of Ohio men did get over the works, but they never came back. At noon Hunter saw that he had too hard a nut for his little army to crack, and secretly gave orders turning back his trains which were far in the rear; but his men fought on, " believing firmly that they were to enter Lynchburg as conquerors if it cost them a week's steady fighting."

From Liberty an excursion of great enjoyment may be made to the top of the Peaks of Otter. We can see the hotel, a white dot, nestling in a cleft at the very top of the sharpest of the twin summits. The view overlooks what is called the Piedmont of Virginia, and can hardly be surpassed. " The lessening hills of the Blue Ridge," said one who knew it well, " with many a lovely valley and brawling stream between, roll downward from our feet, in woody and billowy undulations, ever diminishing until they merge and fade away in the noble champagne country beyond, dotted with still handsome villas and farm-houses. . . . Beautiful sunlight patches floating over the massive and varying verdures of the mountains; clear springs bubbling out from beneath every moss-grown rock; rich flowers shedding brilliancy and perfume even from the topmost cliffs; and dense woods of unmatchable shadow and stateliest growth giving the coolness and repose of perpetual twilight."

The distance to the top is only half a dozen miles, and suitable carriages and drivers can be procured in Liberty at small cost.

The next station of consequence that we rush into, beyond Liberty, is Blue Ridge Springs, a place which looks like a toy town, where the station building is a large hotel. The piazza platform is crowded with belles and beaus, substantial mammas, leisurely papas, and children brown and hearty. Through one set of windows we see the dining-room with laggards at breakfast; through another the office and a billiard table or two; through a third the rich furnishing of a parlor. Down behind the hotel is a deep narrow valley filled with white buildings, great and small, with graveled walks and flower-beds, spring houses and the out-door equipment of a summer resort. These springs are an old-time summering place of high repute medicinally, and higher socially. To the Norfolk people they are particularly well known. Near by, as Baily notes in his big Russia book, are extensive mines of iron ore, which are reached by side-tracks, but no furnaces pollute the sweet, clear air of these charming hills.

Only a little way beyond are Coyner's springs, another watering place where gay girl-faces greet us at the station ; while Gishe's sends a similar bevy in huge sunshades and broad canvas belts to see the train go by. And so, almost before we know it, we leap the current of Tinker creek and roll into Roanoke,—but not to stop, for we are westward bound.

XIII.

WESTWARD BOUND.

The Wheatfields of the Roanoke.—Salem.—Averill's Raid upon the Railway.—The Old Turnpike.—Alleghany Springs.—A Rough Region.—Big Tunnel.—Christiansburg and its Neighborhood.—Central Station.—First View of New River.

HEAD-QUARTERS left behind, we run westward through the rolling bottoms of the Roanoke, crossing many rough and rapid brooks that come down from the hills. The river itself is swift and turbulent, for the descent here is great.

Seven miles from Roanoke stands the old town of Salem, the bulk of which is a mile or so distant. What we can see of it tempts us to alight and discover more. The village lies in a broad valley, is surrounded by large estates, and an air of prosperity and pleasant home-life pervades the whole scene. One of the oldest and most prominent communities in this part of the state, Salem long ago became noted for its highly educated and religious society, which was partly a cause, partly a result, of the location there of two academies of high repute—Roanoke College for boys, and the Hollins Institute for girls. Red Sulphur Springs, nine miles northward and 2,200 feet above the sea, is another point of celebrity in this neighborhood.

This point was reached by the Federal cavalry under Averill, in January of 1864, whose troops were the first blue coats seen in the town. They tore up the railway, and then hastily departed. The ruins remained unrepaired until the following June, when Hunter's army, retreating slowly after the failure to enter Lynchburg through Buford's gap and along the line of the railway, which they still more completely destroyed as they yielded the ground, turned northward from Salem to their last fight on Craig's mountain, just west of Newcastle.

It is with a very smooth and solid track that this destruction of war has been replaced, and the old iron rails, out of which Hunter's men curled such fantastic "neckties" around the oaks, were long ago discarded for steel. The great turn-pike connecting the Valley of Virginia

THE ROANOKE.

with Tennessee, forming a high road through the line of towns between Lynchburg and Knoxville and making an avenue for westward emigration and eastward marketing, is followed closely by the railway, and is seen at the right hand. All along it stand houses, once the homes of the lords of the soil—houses chiefly of brick, with the out-buildings made of hewn logs.

To the left of the train is a confused mass of wooded hills; to the right (northward) the long, straight, sterile Catawba ridge, on whose hither front was old Fort Lewis—a defence of the pioneers against Indians. Then a rough little cultivated nook opens out, and we cross and recross the bright river, which here winds as though it were quite lost among the knolls.

As these knolls lessen in height they become arable, and are all under the plow or in grass, while high up on the tops of the distant ridges great patches of wheat land have been cleared from the forest.

The first stop after Salem is at Big Spring, where there is a group of old-fashioned houses around a gigantic fountain gushing from under a bluff a hundred yards or so south of the track, shaded by noble willows and filled with cresses.

Beyond lie broad and fertile flat lands through which the South fork of the Roanoke comes down from Pilot mountain; and when these are passed the road runs for a long distance upon a high bank, with the river at its wooded base, in and out of rocky cuts, the view (when we can catch glimpses abroad) extending across farm lands to the shapely hills that hem them in with an ever-varying barrier. Thus a plateau is reached where much tobacco grows, and we have climbed 330 feet more to Shawsville, the station for Alleghany Springs four miles distant, whose stages await the train. We did not go over, but Prue read to us about it from a pamphlet, as follows:

"These springs are situated on the head-waters of the Roanoke river, in the county of Montgomery, on the eastern slope of the Alleghany mountains—the most elevated region between the Atlantic ocean and the Rocky mountains."

"That's not quite true," I interrupt, thinking of Cloudland, and several other districts. Then she continues: "The hotel and principal range of cottages occupy smooth and undulating hills, gently sloping to a broad grass-covered lawn of forty acres, extending to the banks of the river. The first panoramic view of the establishment is reviving and refreshing to the dust-covered traveler from the seaboard, and still more so to the feeble invalid escaping from the hot sun of the South. They have the pleasing consciousness, after all the toils and privations of travel, that they have at length reached a spot where ease, comfort and repose await them. This feeling is the first step toward the restoration to health. . . . The accommodations are first-class, and afford every convenience and comfort both to the invalid and the pleasure-seeker. Pure spring water is conveyed by pipes from the mountains to every part of the establishment, and guests are supplied with hot, cold and plunge baths."

The hills rise steep and irregular all about us as we proceed; brushy to the top, but showing few large trees, for these have been cut out as fast as they became available. Now and then a fine house will stand in some more open spot, for this red soil is rich where enough of it can be found together, but the principal signs of habitation appear in the shape of rude cabins in the midst of small neglected clearings. We see no sheep, but know they belong here—as the Arab in the fable knew a camel had passed his tent—by the tracks of their nimble feet on the steep gulchsides.

At Big Tunnel—which is great only by comparison, and takes but a minute or so to traverse—is another little station, where "out in its beautiful nest, only a mile from the road, sits the famous Montgomery White Sulphur Springs." It was from these heights, looking over the path we have come, that Edward King, author of "The Great South," "came suddenly upon the delicious expanse of the Roanoke valley, bathed in the splendid shimmer of an afternoon autumn sun, and fading into delicatest colored shadows where the mountains rose gently, as if loth to leave the lovely and lowly retreat. The vale was filled with wheat and corn fields, and with perfect meadows, through which ran little brooks gleaming in the sun."

That epitomizes the view very nicely.

Still ascending, we presently reach the top of the pass over the "Alleghany mountains" (by which obscure term map-makers trace the divide between the Atlantic and the Mississippi) at Christiansburg, another old time village, where a highway north and south crosses the east and west turnpike. These roads penetrate a fruitful and beautiful region to the northward, and lead to such points of special interest as the Yellow Sulphur Springs and Blacksburg Agricultural College. In regard to the former, Prue's guide-book informs us that it is three and a half miles distant and 2,200 feet altitude. "The mineral properties of the water are well known, and have been tested by thousands, and invariably with good effect. It is alterative and tonic in its action, possessing all the qualities which are usually found in the best alum springs." Central, the next station, is a half-way point between Lynchburg and Bristol, and hence chosen by the railway as a place for repair shops, a round house and other works. The Union cavalry, during Averill's remarkable raid in the spring of 1864, reached this place and destroyed track and buildings; a house near the station shows yet a perforation made by one of the shells.

From Christiansburg, which is 2,000 feet above the sea, we have been running downward, past the coal-bearing Price mountain on the right, through a very picturesque region, and at Central come suddenly out upon a broad river, curving grandly about the great green hills that guide its course. It is the New river, and presently crossed upon a lofty iron bridge, from which a magnificent picture is presented in each direction.

The New river is now one of the *oldest* rivers in the Union, so far as knowledge of it goes. Rising from springs on Grandfather, in the Iron Mountain range of North Carolina, it pursues a great curve toward the east through the serried uplands of the Virginian Appalachia, and empties into the Ohio opposite Gallipolis, almost due north of its rising. Beginning as the "New," it changes its name below the entrance of the Gauley river, in West Virginia, and becomes the "Kanawa."

Here at Central the stream is quiet, willow-fringed and bordered by farms, save where rocky bluffs approach too closely. Just beyond the bridge is the station of New River, whence departs the branch line running down the river into the coal regions of the Bluestone valley and Flat-top mountain, the centre of which is at the town of Pocahontas, some seventy-five miles north-westward. This branch line pursues for a long distance the course by which the stream has cut its way athwart the mountain ranges, and affords the traveler a sight of one of the most interesting and beautiful cañons in the country.

XIV.
NEW RIVER CAÑON AND MOUNTAIN LAKE.

Cloyd's Mountain.—The River Gate-way.—Cliffs, Bastions and Pinnacles.—Luxuriance of Foliage.—The Narrows.—Pocahontas and its Coal Mines.—Eggleston's.—The Road to Mountain Lake.—Sunset from Bald Knob.—Boating upon the Lake.—The Glories of the Forest.—Prue's Summer House.

UPON leaving New River station the railway crosses the peninsula enclosed in the great Horseshoe bend, but, half a dozen miles ahead, again approaches the stream where it attacks its first obstacle, now divided, by its success in forcing a passage through, into Brush mountain on the east and Cloyd's on this, the western side.

This first gateway is an impressive suggestion of what will follow. On each side the mountain stands with its feet laved by the current, and at a low stage of the water the eye can trace at a glance the continuous ledges of steeply inclined strata which match one another in the opposite headlands, and are connected across the bed of the river. The mountain is densely timbered, but through the trees and bushes these great slanting ledges protrude in tilted shelves, whose slope corresponds precisely with the southern face of the ridge. The roadway has been dug out of the steep hillside, and everywhere on the inside is a rocky wall, sometimes breaking down where a ravine has worn a hollow; sometimes rising many feet sheer above us; sometimes so hollowed under that great masses overhang the cars.

The river here cuts squarely across the range, and we can see,—but not count, for they are too many—the upturned, black edges of the eroded strata stretched across the streamlike miniature dams. But soon the river changes its course, and then these parallel lines run diagonally across it, or even lie lengthwise of its current in the elbow of some bend. Wherever that is the case the river narrows and deepens, because its eroding force, applied lengthwise the outcrop, would act more forcibly to cut downward than sideways; while for the opposite reason, the stream is always broadest where the stratum edges run most nearly transverse to its current. Over these submerged, or half submerged, ledges the water passes with much commotion, and sometimes, where an outcrop is harder than its fellows, and therefore more prominent, there will be a little cascade.

This Cloyd's mountain overhead is the same at whose western end, where the highway passes from Newbern to Pearisburg, was fought the terrific battle between the Union forces under Crook, and the Confederates under Jones and Jenkins, in which Brig. Gen. R. B. Hayes (afterward President of the United States) was one of the commanders. The Rebels were entrenched in a strong position upon the mountain, but at a terrible cost of life were at last driven out, and pursued with constant fighting to Newbern, where the destruction of railway bridges and stores, which had been the object of the expedition, was effected.

The cliffs, under which we are passing, are very beautiful. The new rock, white or yellow, contrasting finely with the hoary gray of the natural exposures, and overhung by the dense and varied foliage of the universal forests, presents some novel combination of form and color every moment, for we are continually changing our point of view in following the windings of the gorge. Sometimes the cliff wall and its sloping mountain-cap, a bit of river, a fragment of corn field and an uncertain background of far away highlands, are held for an instant in the frame of deep cutting; then a shoulder of rock, or a grove of shining maples, slips between and blots it out, but with the same motion places upon the screen another picture equally enticing yet wholly unlike in its composition. With such elements at command, what cannot be done?

At Staytide station (where a gay party get off, bound for Eggleston's and Mountain Lake), Cloyd's mountain has been passed, but opposite stands the still more lofty barrier of Walker's mountain.

Emerging from a short tunnel just below Staytide, we are confronted on the opposite side of the river by vertical cliffs of rocks that are broken into bastions and pinnacles in some parts, in others remain massive and bold. They rise several hundred feet, straight from their reflections in the oily flood so deep and still at their feet, and bear upon their shoulders the rugged summits of Gap mountain. On this side, also, sandstone cliffs surprise our eyes by their height. We scarcely expect such greatness in southern scenery as is here to be found. Assailed by blasting, to furnish a passage for the railway, where otherwise no room for a track existed between the cliffs and the water, the strata have so broken apart as to leave great square corners, with protruding points and ledges which cast bold shadows. Nor are the rocks white and garish, but softly colored in browns, pale yellow and dead red tints; while above and beside the new exposures rise the jagged precipices shaped slowly by water and air, and dark with the storms of unnumbered seasons. They are reared above the trees in grand pyramids and towers, or jut out like the prow of some huge vessel, or stand in thin protruding walls set edgewise into the verdure-hidden hill,—for everywhere, you must remember,

" The scarred summit's rifted seams
Are bright with glistening pines."

It is the exceeding luxuriance and diversity of the foliage, and the beautiful way in which these hoary and massive old rocks are embowered in trees, shrubbery, vines, herbage, moss and lichens, which make them so picturesque and prevent that feeling of roughness and sterility that becomes oppressive in the far West or among the Canadian hills.

Thus we wind with the urgent river through the Narrows, where

NEW RIVER SCENERY.
[From Photographs by C. H. JAMES, Philadelphia, Pa.]

Pearis, Wolf Creek and the East River ranges crowd close together on the west, and the huge form of Peter's mountain towers massive on the east. But the river long ago won its right of way between, and we follow, unable to see out, and undesirous to do so, for the "storm-tossed Titans" around us, and the restless pushing stream, are enough to fill our eyes and minds.

Clear of this montanic labyrinth (it is wonderful to look back at the huge wall and try to understand how you squirmed through) the road leaves the New and takes its course up East river, which flows along the northern base of the range of the same name, with Black Oak mountain opposite. This gorge is narrow, abrupt or often craggy on its sides, forested from top to bottom, and hence most romantically wild. Saw mills are being put up all along it since the advent of the railway, the logs to supply which are sent rolling and sliding down long paths cleared on the faces of the high, steep hills, which press upon us with rocky walls almost unclimbable, every dark old ledge just now lighted up with thickets of purple poke-berries. Oakvale is the railway station for Princeton, the seat of Mercer county, West Virginia; and Winonah that for Pearisburg, county seat of Giles, Virginia. At Graham's we have reached the head of East river, and thence descend along one of the confluents of the Bluestone to Pocahontas, through a gorge so narrow, and occupied by so tortuous a stream, that it is crossed seventeen times in the nine miles.

Pocahontas is a rough, new town of frame houses, nearly all of the same pattern, built by the company whose coal mines are there, and painted in the most original colors possible for the mind of man to invent and combine! In order to get room for a town, the forest was cut away from a hill side—the expansion of the bottom of the gorge there being none too great to accommodate the many railway tracks, coal-coking ovens, and company's shops of one kind or another. In the midst of the stumps and boulders of this steep hillside the houses were set down in streets which some day, perhaps, will be terraces. The place looks just like one of the new coal or silver camps in the far West, and strongly reminded me in particular of Carbonado, in Washington Territory. Pocahontas will grow better, however, when she has had time. There are churches and school-houses, a club and reading-room, good stores and the beginning of a prosperous town; but never, I venture to think, will it be a delightful place.

Reversing each picture as we descend the railway along New river, the "lofty and luminous summits" take on a new aspect and beauty quite as novel and enchanting as when we admired them going up.

It seems a short trip, therefore, that brings us to Staytide, where we disembark and are ferried over to the New River White Sulphur Springs, or, as the place is more generally called, after the owner of the estate— "Eggleston's."

At Eggleston's, one finds a long rambling old house with a broad lawn and grand trees in front, an orchard, garden and sunny hillside behind, standing on the river bluff where it commands a view of the bend and both shores. Old in tradition, as in architecture, fashion is here left behind, and health, fun and comfort reign. For him who enjoys the water—and who does not?—there is the broad, deep and placid river, where he may pole a punt, or paddle a canoe, or anchor and fish, or find a shaded nook and swim in the cleanest of floods. For the lover of scenery, there is the broken front of the mountain, carved by nature's deft chisels into a thousand buttresses, arches, and pinnacles, half veiled in clinging verdure and the nestling place of playful sunbeams and coy shadows. For the enthusiast in natural history or art, I know no more fruitful spot. For the invalid, there are peace and beauty more healing than the most beneficent waters.

Eggleston's is now the point of access to Mountain Lake. This is a body of water a mile in length, poised near the summit of a spur of the Alleghanies at an elevation of 4,500 feet above the sea. A century ago it was much smaller than now (a story that formerly no water at all was there, save a rill, is unworthy of belief) and became a favorite place for the graziers to collect and salt their cattle. Hence it came to be called the Salt Pond, and the lofty semicircular ridge that surrounds it was named Salt Pond mountain, long ago resorted to by picnic parties, who reached it by the good road which crosses over the range from Christiansburg and Newport to Union.

The road from Eggleston's winds upward along creek gorges, through a very wild and lonely region, where the mountaineers are content with wresting small crops of corn and tobacco from the rocky soil. The streams occupy gorges which in summer go nearly or quite dry, yet are liable to become the sudden conduits of a thunderstorm, when their channels will be filled with a raging flood, no trace of which remains next morning. As you approach the summit, wide views open out southward and eastward which gradually expand as each succeeding ridge-top is surmounted and the way grows steeper. Going up in the afternoon, you do not halt at the lake, but keep on to Bald Knob, the highest point of the Salt Pond mountain, whose apex is so beaten by gales and burdened by snow that nothing better than stunted bushes has been able to take root amid the clefts of the rocks. Here you arrive at sunset, and in its clear light your eye can sweep a circle which passes through the boundaries of five states. There is nothing to show where these boundaries ie, however, and pride of locality is rebuked as one sees how homogeneous is the whole landscape.

It is a very wrinkled, disjointed and savage world that you look upon, too. In every direction as far as sight goes, from the azure crest of Kentucky's Cumberland on the west to the ragged horizon behind the twin peaks of Otter on the east, and from the magnificent mountains of North Carolina, where the Kanawa takes its rise, to the faint West Virginian sky-line where it fights its troublous way toward the Ohio, all is mountain and valley. Rank behind rank, ridge undulating with ridge, peak rivaling peak, spur flanking spur, color answering to color in the regular gradation of distance, which ever way you gaze the magnificent picture awes you with its breadth and weight and sublime repose. It is a vast harmony in blue; the higher lights of the elevations softened and the shadows in the depressions illumined by the haze which veils, subdues and idealizes, lending to a stern, rough world the hue of

> "——the clear and crystalline heaven,
> Like the protecting hand of God inverted above them."

How can one portray a scene like that, or tell the emotions?—the scientific satisfaction with which a map-maker and geologist would scan the wide area beneath his eye, tracing the system concealed beneath the seeming disorder; the watchfulness with which a poet would study the delicate blending and comparison of colors, all graded to purple and blue with infinite taste; the ecstasy of the poet, the veneration of the seer?

Then as the orb of day drops slowly into the smoky atmosphere along the western horizon, to illumine it with a red and coppery light momently changing and flooding that proud half of the world with golden mist, while behind us the ridges grow dark and are massed together in purple gloom—how can any one describe *that* either; or make you understand the play of dazzling color shooting from behind the serrated wall so sharply marked upon the brilliant sky?

> " Beside us, purple-zoned, Wachusett laid
> His head against the West; whose warm light made
> His aureole; and o'er him sharp and clear,
> Like a shaft of lightning in mid-launching stayed,
> A single level cloud line, shone upon
> By the fierce glances of the sunken sun,
> Menaced the darkness with its golden spear."

The lake (I regret that the historically suggestive name Salt Pond, has been thrown out) is reniform in outline and a mile in length. At one end a large clearing has been made, and here stand the various buildings of the unpretentious but comfortable hotel, which include near the water's edge a boat-house, bowling alley and billiard room. There are also a few private cottages. Taking a boat we slowly row down the lake, and enjoy its strange and inspiring beauty—the interest of solitude and wildness. When your back is turned to the hotel, nothing is to be seen but a forest of remarkably varied growth, rising amphitheatre-like and unbroken from the water to the crest of the lofty ridges. At the lower end of the lake we land and climb among broken crags to where a summer-house is perched high among the trees, and haunted by birds.

There is a cascade and creek gorge of special charm not far away, and two or three look-outs on the mountain from which ample landscapes may be seen; a ramble of paths through the forest; a remarkable cold spring; and always the sunrise and sunset from Bald Knob. But Prue and I think that, if ever we are permitted by happy fortune to spend a month at Mountain Lake, it will be to the sequestered summer-house among the mossy rocks and sunlit woods at the lower end of the pond that we shall oftenest go.

XV.
THROUGH SOUTH-WEST VIRGINIA.

New River to the Tennessee Line.—Dublin.—Mines and Furnaces of Cripple Creek.—Martin's, and the pretty Maple Shade Inn.—Max Meadows.—Wytheville.—Rural Retreat and Marion.—Bass Fishing and Mineral Springs.—Glade Spring and Emory College.—Trout.—Saltville and its War History.—The Battle at Wytheville.—Abingdon and its "Girl Graduates."—Bristol and Johnson City.

FROM New River westward is a charming ride among the mountains. Indeed, the great drawback to the pleasure of traveling amid these superb fastnesses of nature, is the singular one that the eye and brain are sated by the constant succession of noble outlooks which occur as we progress. Here the titans seem to have vied in their gigantic upbuilding of cliff and pinnacle. Here the tempest has its lurking place, and the fountain heads of mountain torrents spring forth gladly, and go dashing down their rocky beds to the far away alluvial valleys where mankind shall chain them, and make them turn huge mill wheels, stain their purity with the outflow of factories, and send them on to lose themselves in the salt bays that reach up to grasp them from the wide and restless sea. We look up and about us with awe and reverence, and even the irrepressible Baily for once is hushed.

Dublin, where a fight followed that on Cloyd's mountain, having been

THE MAPLE SHADE INN.

passed, we skirt the rough declivities of Peak Creek Knob and Draper's mountain, with Little Walker's mountain filling the northern horizon some miles away. At Martin's, in a comfortable little basin among the hills, we enter the mining regions of south-west Virginia—a region whose almost unlimited resources in this direction have hardly begun to be developed yet. Close by are the Bertha zinc works, both the mines and furnaces of which are well worth visiting. Northward a branch railroad, nine miles long, leads to anthracite coal mines in Pulaski county, while a much longer railroad nearly completed goes southward along Cripple creek, in Wythe county, to a larger number of mines and furnaces, producing iron, lead, zinc, copper and coal. At present the ores and product of these mines and furnaces (its pig is said to have been sold for the highest price any pig iron ever brought in the world) is taken for shipment to several small stations along this part of the line, of which the chief is Wytheville, the county seat.

No man who is interested in minerals and their reduction should fail to stop and inspect this remarkable district ; and in order that visitors may do so with comfort, a fine new hotel, like Luray Inn and the Hotel Roanoke, will have been opened at Martin's before this page gets into type, called the Maple Shade Inn. Following in its architecture the favorite models of the "early English" school, modern and luxurious in all its appointments, managed by the most experienced northern hands, this fine hostelry will make a halting place of peculiar attractiveness, not only to those whose business or curiosity in mining matters leads them to stop there, but to the great tide of pleasure travel which drifts ceaselessly up and down this favored region, and is glad to pull up, where it can be sure of creature comforts, long enough to get more than a passing glimpse of the mountain scenery. To persons coming from the southern lowlands, to whom the coolness and northerly beauty of these ranges and headlong brooks are especially grateful and fascinating, the establishment of this hotel will be particularly opportune.

Just beyond Martin's is Max-Meadows, an ancient lake basin, where the wide valley presents a lovely pastoral landscape, in which the furnace stack and ore-wagon seem anomalous.

"What are you saying, Prue? That your guide-book hits it off neatly? Let us hear,

' And lend to the rhyme of the poet
The beauty of thy voice.' "

" That is asking a good deal, but here is the description :"

Groups of cattle are scattered about, shading themselves under the trees or nipping the succulent grass. The river here is broad and clear, mirroring in its placid breast the verdure-bordered banks, to whose sides the railway confidingly clings. Amid the gently sloping hills, this little meadow-town looks quietly out on the world ; and the busy men, who handle great loads of iron and great bags of shot, from the iron and lead mines near, do not seem to realize that a *scribe* is taking note of them, and that posterity will read of their enterprise in type of emulating character.

"That's good. What is written of Wytheville, which you can't see very well, because it is a mile or so from the station ? I never went up to the town, but I know you get a stunning good meal at the station, in the queer basement of Boyd's big brick hotel, when your train happens to hit it at the right hour."

"Well, the book speaks twice of the good fare, once of the good society and extols the cheapness and excellence of the livery," says Prue glancing along the page.

"That's a strong point, ' Baily adds, "because Wytheville is a summer resort, and one of its great attractions is the good fishing to be had in the neighboring mountains, and the remarkably interesting scenery their gorges and summits contain. To enjoy both sport and sight-seeing one needs serviceable horses and vehicles, and doesn't want to pay extortionately, so I score one for Wytheville on that. Shall go there myself next year. What else Madame Prue?"

"Besides the naturally attractive features of this place and its surroundings," she resumes, "it is grading and paving its broad streets and has brought into the heart of the town the waters of a fine alum, sulphur and chalybeate spring through a system of pipes—perhaps the only improvement of the kind in this country. On the next page it mentions that Sharon Alum and Chalybeate Springs, in Bland county, is only eighteen miles from Wytheville, in the midst of a pretty country containing good shooting and fishing."

Stations pass rapidly. Crocketts, Rural Retreat—the highest point on the Norfolk and Western Railway—Atkins and Marion are all of more importance to the freight than the passenger department of the railway, for they are points of shipment of ore and pig metal of iron and copper, some of which is brought from North Carolina.

The last named is a village of some size, and contains a boys' academy and a female seminary. The middle fork of the Holston river runs through the town, and is full of bass; while the mountains southward—particularly White Top—are a paradise for fly-fishers in search of trout. Not far away are the Chilhowee, White, Black and Red Sulphur springs, the water of each one of which is famous for its curative power over scrofula and other ailments.

Just below is Glade Spring depot, about which the Madame reads:

"Glade Springs is a little village from which the tourist can reach many points of interest. Washington Springs nestles among the hills two miles away. The Seven Springs, noted for the 'Seven Springs iron and alum mass,' are two miles out in another direction. White Top mountain, noted for its bears and speckled trout, looks from a distance down on the village streets. From this place a branch road, ten miles long, leads out to Saltville. . . . Nature has done a great deal for the Old Dominion, but with characteristic energy these south-western Virginians have had to 'have a hand in it.' They have put churches in the groves, mills on the streams, barns in the valleys, colleges on the hills. They have not always improved upon Nature's work, but in many instances they have not marred it. Here, two miles from Glade Springs, is a pretty valley with a high hill in the centre, and on the top of this, as if to proclaim to the world the appropriateness of its motto:

'Mens sana, in corpore sano,'

Emory and Henry College is built."

"Strikes me they might have found space for something more about Saltville," Baily growled.

"Do *you* know anything about it?" I asked.

"Quantities of facts," he declared, opening the red book. "Saltville is the centre of a natural basin or valley, which is one of the loveliest spots in Virginia, and that is about equal to saying, 'in the world.' In the centre are springs of saline water and gypsum mines, and here are extensive salt-making works. It was upon these works that the Confederates depended almost wholly for salt during the war, and from the very start the Union generals schemed to destroy them; but they were too far within Secessia and too well guarded. In the spring of 1864, however, the Federal department of West Virginia sent a strong cavalry expedition to work all the ruin it could along this railway, and while Crook was to operate at New River, as we have already learned, to Averill was assigned the attempt against Saltville. But when Averill got into Tazewell county—and a sweet time he must have had of it among those complicated ridges!—he heard that the defences at Saltville were too strong for him to attack as he had no artillery, so he turned against the bullet-making lead works at Wytheville; but General John Morgan moved his troops and guns at once from Saltville to Wytheville, and fought a battle which Averill got so sick of, that during the ensuing night he decamped eastward, and contented himself with wrecking the railway and shops near Christiansburg. Morgan's men went on by train to Dublin, but were too late to do more than cover the retreat of their comrades who had been beaten by Crook at Cloyd's mountain and Newbern."

Abingdon, which the brakeman called about the time Baily shut up his book, revived in my mind the story of Daniel Boone, whose trail to Kentucky ran this way; then Prue said she had a friend who went to school here, and so knew that the pleasant town had several girl's boarding schools, of high repute in the far South, whence parents are wisely fond of sending their children into these healthful highlands to pass their school days. Emory College is near enough to be convenient for flirtation purposes, and the pretty maidens in their walks abroad can usually get a glimpse of stalwart sophomores, or even capture, now and then, the matured heart of a grave and reverend senior. Abingdon is a flourishing place of a couple of thousand people, beautifully situated, and surrounded by rich magnetic ores of iron, and by valuable beds of variegated marble.

A few moments more and we are at Bristol, the terminus of the great Norfolk & Western Railway, and on the boundary of the Old Dominion, for half the town is in Tennessee and half in Virginia.

"Bristol," reads the Madame from the last page of her guide-book, "is the point of junction with the new railroad to be built through Scott county to Cumberland gap, in the direction of Kentucky—the centre of another district which must become noted as a mining and manufacturing section. Scott county has thick beds of the finest Tennessee marble, iron ores and coal. Lee county, a short distance further, has extensive deposits of the famous red hematite iron ore; and when all these ores are brought by the new railroad into communication, at Bristol, with the magnetic and the brown iron ores of Washington and Johnson counties, great furnaces for their reduction must be built at Bristol. Bristol has already given evidence of decided improvement and extension, and is destined to be one of the most prosperous and distinguished of inland towns."

We are hurrying to a delight ahead, and do not stop at Bristol. This trip is not in search of commercial gain, but mental relaxation in the enjoyment of the beautiful and the ennobling in nature:—

> "Absence of occupation is not rest;
> A mind quite vacant is a mind distressed."

So we ran straight on to Johnson City, Tennessee, and put up for the night in a hotel, which, if not gorgeous, was at least comfortable.

XVI.

ROAN MOUNTAIN AND THE CAÑONS OF DOE RIVER.

A Narrow Gauge Side-trip.—The Cranberry Iron Mines.—Elizabethtown.—Approaching the Canon.—Fierce Torrents, Lofty Headlands and Flower-hung Precipices.—The Road to Roan Mountain.—Peculiarities of the Forest.—"Cloudland."—Above a Thunderstorm.—The Hotel on the Peak.—Exploration, Sport, Science and Love-making.

OUR halt at Johnson City was to be prepared to spend the next few days in "Cloudland,'—a region justly so-called, since it lies six thousand feet above the level of sea, supported upon the Atlas shoulders of Roan mountain. It is reached by the East Tennessee & North Carolina Railway (narrow gauge), which unites with the main line at this junction, and is carried along the gorge of Doe river. Its trains are conveniently arranged both for those who wish simply to see the gorge at a cost of no more than half a day's time, and for the happy others who may climb Roan mountain. For a tourist to miss this trip, I assured my party, would be one of the greatest errors conceivable.

The former (short-trip) class of travelers can go up as far as Cranberry station in the morning, get dinner, and a ramble at the fine hotel maintained there by the iron mining company, and return in the afternoon to Johnson City in time for the evening trains east and west.

Besides the scenery of the cañon they can visit the mines at Cranberry, where, by means of level tunnels piercing the hillside, is obtained a bluish, magnetic ore of iron of peculiar purity and unequaled in the United States, Baily assures me, for the making of Bessemer steel. The mining operations, only recently begun, are steadily enlarging, and pig is made on the spot by the charcoal process, wood being cheap and plentiful in the neighborhood.

The chief interest of the Doe River trip, however, lies in the passage through the cañon, which recalls to me more nearly the appearance of Grape creek, near Cañon City, Colorado, than any other gorge I know.

The Doe river is a stream as broad as a village street, which rises high up in the Roan Mountain range, and empties into the Wautauga. Though its later progress is quiet and dignified, as befits the serene ending of a busy youth, all its early career is a headlong race through the wilderness, and its history is that of one "who overcometh."

On leaving Johnson City the first ten miles of the journey thither carries one through a rich farming region, affording a most excellent sample of the rural appearances, population and picturesqueness of *old East Tennessee. On the south towers a double mountain—a spur of the Unaka range, one part of which is locally called Brier mountain, while the further crest is named Buffalo, because of the "hump upon his back."

Ten miles from Johnson City is Elizabethtown, the county seat of Carter, or *Keeyarter*, as a northern ear will interpret the gliding pronunciation inherited by these people from their Scotch-Irish ancestors. It is a pleasant village in the centre of a plain filled with fertile fields and surrounded by shapely and richly-tinted hills; and the copper cupola of

ON DOE RIVER.

its court house stands like an accent-mark above the brown roofs, to emphasize it in our recollections.

Only a little way farther on the close-crowded foothills, standing like outer works of a grand fortress, rise compact and precipitous, and out of them gushes the noisy and vehement river. The track seeks its margin, where it is laid along scarps carved from the solid mountainside, or upon a bed of broken stone thrown down at the edge of the stream. The river comes tumbling towards us with madder haste and whiter foaming the farther we ascend, for its path grows more steep, tortuous and obstructed, until it can only be described as one long rock-tormented cataract—white and yeasty where it struggles fiercely in the rough or narrow places ; glassy where in even flood it curves smoothly over some ledge or bowlder ; rich luminous green in shadowy pools, whence bubbles rise like buoyant diamonds to disappear in twinkling sparkles of colored light. Plants creep down as near to the rocky channel as they can, fearless of the floods ; and shrubs that miraculously have taken root hold sturdily to their place beside the water, cooled by its flying spray. Above all, on each side, rise the rugged, sliding, eagle-haunted, thickly wooded steeps, to their pinnacle-studded crests poised a quarter of a mile in air—

"A wild and broken landscape, spiked with firs."

Yet this is only the beginning of the gorges. While we are sweeping round its sudden curves, threading brief tunnels, or skirting bold headlands from one *cul de sac* to another, our attention engrossed by the river and the grandeur of the primeval hills, we suddenly find the wooded gorge become a rocky chasm, and that, with one foot in the boiling torrent, our venturesome train is pushing between vast walls of massive granite and gneiss that rise hundreds of feet straight overhead—sheer as a plummet's fall and so lofty that the great spruces perched dizzily upon their verge seem the merest shrubs. These cliffs do not stand flat and unbroken like an artificial wall, however, but have been worn into a series of mighty headlands which we review at various angles, now near, now remote—gazing ahead at the towering height which we tremble to approach, straining our necks in vain endeavor to scan its top while we are close underneath, looking back at it in surprised delight as it poses in splendid dignity behind our departing train, or salutes its royal brother across the gorge.

Prue and I have seen far taller cliffs, and a hundred miles of them where here is only one. We have run for half a day amid the ringing terrors of the Royal Gorge and the Black Cañon, with walls so near we could almost touch them, and half a mile in altitude. The face of none of these Doe River cliffs, perhaps, is more than six or eight hundred feet in vertical height, capped here and there by a spire or pinnacle, and they form, as I have said, a series of prominences, rather than a continuous wall ; but we decided that there was more of charm for the eye and satisfaction to the heart here than among the greater glories of the Rockies. The comparison, nevertheless, is not quite fair, though it is sure to be made by all travelers who have seen both; for in place of the purity of the air and the nakedness which magnify the impression of vast size and distance in the far West, we have here a soft atmosphere, yellow light, and the mantling beauty of diverse vegetation which belong to the southern climate.

The arrangement of the rocks and their varying hues, painted by the weather, contribute greatly to the fine effect, too. The strata are not horizontal, making every line either level or upright, but they slope steeply down stream, and the shape of every hill-top, headland, river-fall and horizon line, conforms to this pleasing angle. The whole world hereabouts, as Prue puts it, is "cut bias." Athwart the fronts of these gigantic cliffs run bands of vari-tinted strata, separated by black lines and great protruding ledges, which carry slanting ranks of trees, one above the other, their backs set close against the cliff, their branches all reaching fearlessly forward over the abyss. Everywhere upon the lesser ledges, and upon all the jutting points and pediments, and everywhere preserving and emphasizing the graceful inclination of the rock-layers, are rooted flowers, ferns, chevrons of emerald grass, bristling evergreens and gnarled dwarfs of other trees, from the foam-splashed foundation to the towers that challenge the thunderbolt and catch the first flush of dawn. The cliffs, then, are grand, not only, because of their height, massive breadth and unswaying solidity as they lean over the chasm, but also are rich in beauty.

Such, feebly rendered, was the impression this wonderful gateway through the mountains made upon us. Some day a greater leisure will let painter and poet study it, and then, I hope, a far better hand than mine may portray it for you as it should be done. Why, my dear reader, do you not go there and try to do so yourself?

At Roan Mountain station comfortable hacks are in waiting to carry the traveler to the summit, a dozen miles distant. The road is a good one—not, of course, a macadamized boulevard, but much better than one could expect ; and it runs almost uninterruptedly through the dense forest, clothing the foothills. The botanist observes changes as the grade rises; plants of the lowlands disappearing one by one, and varieties of trees and foliage presenting themselves which only belong to high levels. At last almost nothing in the shape of trees, except the balsam fir, can be seen, denoting that the summit is near ; and when the top is finally attained you find it altogether bare of wood. The unshod pate of the mountain, nevertheless, is carpeted by a turf of luxurious grass, variegated in summer with innumerable alpine blossoms small in size, and sitting close to the ground, but lovely beyond their more rank and showy sisters of the lowlands.

The few acres of open land on the apex of this huge elevation (its neighbor, Mount Mitchell, is the loftiest point east of the Mississippi) is far above the range of ordinary storm-clouds, so that the novel spectacle of thunder, lightning and rain a thousand feet below you may often be witnessed. It is called "Cloudland," and furnishes the site of a hotel, made of logs, where sixty or seventy guests can be accommodated in great comfort, and even a hundred have been stowed away. "It is never warm up here, and people are willing to sleep close together," it was explained to us. Plain, but wholesome and satisfactory, fare is furnished, and the charge at present is only two dollars a day.

Though here the poet's wish for "a lodge in some vast wilderness" would seem to be realized, since there is no sign of humanity beyond the hotel clearing, ample opportunities for amusement and pastime exist. The loquacious landlord will tell you of many a point of lookout to be visited, one after another, each spreading beneath you a new landscape "wide, wild and open to the air," yet made up of the same glorious elements that constitute the others. There are caverns to explore ; bee-trees to search out ; glens to lunch and dream in, lying beside babbling brooks upon springy cushions of fern and moss ; crags to climb, and precipices to try the nerves of the most dauntless ; while every stream that gathers headway down the mountain is haunted by trout, and each grove tempts the sportsman with certainty of small game and the alluring chance of a deer or bear or wildcat. As for flirtations—given the girl of your heart, and the whole world has not a more inspiring spot ! while the sober minded scientist, turning his back on such immaterial frivolities as games on a hotel lawn or love-making excursions to crags and crannies, can find here a Labrador brought south for his study, since the great altitude of this peak makes its climatic conditions, and consequently its fauna and flora, almost arctic. For old and young, grave and gay, therefore, "Cloudland" furnishes peculiarities of occupation and novelty in entertainment, such as is combined no where else that I know of in this country.

XVII.

THROUGH EAST TENNESSEE.

The Great Appalachian Valley.—Early Settlement of Tennessee.—The Home of Andrew Johnson.—Rogersville, Morristown and the Marble Quarries.—The French Broad.—Asheville and the Highest Mountains.—Warm Springs.— The Holston River.—Knoxville.—Rich Landscapes.—En Route to Chattanooga.—Athens and Cleveland.— A Railway Centre.—Chattanooga and Its Wonderful Progress.

THE space and purposes of this little book will permit only a hasty review of what we saw and did on our way through Tennessee and the states farther South. So hasty, in fact, that I think it will be best to bid farewell to the gentle Prue and the gay Baily,

" 'Tis better to have loved and lost
Than never to have loved at all ;"

and to speak hereafter in general terms of what the continuance of this excursion, and subsequent winter travels, taught me to regard as the most interesting routes of travel to and through the Gulf States of the Union.

The East Tennessee, Virginia and Georgia Railroad, whose eastern terminus is at Bristol, is a legitimate part of the railway system we have been pursuing, since it follows that continuation of the Great Appalachian Valley which lies between the highlands of western North Carolina and the Cumberland mountains, and forms East Tennessee. It is a continuous avenue between the mountain ranges and almost a direct line of railway from Harrisburg, Pennsylvania, to Chattanooga, Tennessee.

East Tenessee presents many points of interest to the farmer, the lover of out-door pictures, and the student of human nature. At Jonesboro, just beyond Johnson City, was the first settlement in the state; a company of Scotch-Irish immigrants to North Carolina having struggled through the mountains and driven their stakes in that locality, where they were soon joined by pioneers of German descent from Virginia. Wonderful views of the mighty mountains which form the source of so many great radiating rivers are caught southward across a rich and thickly occupied region, as we push on toward Greenville, the most important town, at the eastern end of the state. To Tennesseeans its history is linked with many memorable names and events; but to the world generally this pretty town is noteworthy chiefly as the home of Andrew Johnson, whose house is out of sight, but whose monument is conspicuous upon a

ALONG THE UPPER FRENCH BROAD.

hill-top east of the village and close to the track. At Rogersville Junction is a station eating-house—there is one also at Jonesborough which I forgot to mention in passing; and both can be most heartily recommended. The branch road which comes in here is from Rogersville, an agricultural and academical centre some fifteen miles to the north, in the midst of quarries of the celebrated "Tennessee" fossiliferous marble, whose mottled-brown color makes it so handsome and valuable. At several stations between here and Morristown, the freight platforms are heaped with blocks of this marble, which occurs widely throughout Hawkins county.

Morristown has two thousand inhabitants, and promises to become a place of great importance. Here branches the railway into North Carolina, which, crossing to the French Broad at the mouth of the Nolichucky, follows the former stream up into the heart of the mountains that rear their magnificent forms against the southern horizon. Let us diverge upon it for a moment.

Perhaps no river in the country is invested with more romance, and is more highly worthy of its reputation for beauty, than the French Broad. Born in the Swannanoa gap, its childish cascades leap and prattle through fragrant thickets of rhododendron and azalea, whence it rushes with fast-gained strength for miles and miles among the loftiest mountains on the Atlantic slope. Many a talented pen (especially Christian Reid s) has helped make famous this glorious region ; and it is hardly necessary for me, at this late day, to more than mention the names *Asheville* and *Warm Springs*, to fully remind all travelers of the comforts that await them at the several centers of traffic and repose.

Asheville (reached by the Western North Carolina Railway, connecting with the Morristown Branch of the East Tennessee, Virginia and Georgia), is the place from which most excursion parties take their departure for the exploration of the mountain region. Within a distance practicable to walkers, or suitable to a morning's drive or ride upon horseback, are a long list of eminences, each with its special adventures and view.

Varying this, charming nooks and fishing resorts occur along the river ; alpine glens and ferneries ; and winding country roads, where the queerest people and the most primitive home-life may be studied. From the greater heights, like Pisgah, which are the object of more extended, yet easy excursions, you may count scores of peaks exceeding six thousand feet in height, and countless more approaching it. The ascent of some of these is an easy matter; to climb others becomes a teat worthy of the Alpine Club, and one which stimulates many a laggard ambition to unwonted effort and proud success. Branches of the Western North Carolina Railway, as well as carriage roads and bridle paths, give access to most of the prominent peaks, including Mitchell, the captain of them all.

"Within a day's journey on horseback, are some of the finest fishing streams in these mountains—Doe river and its branches in the Black mountains, the east fork of Pigeon river, and still further on, the beautiful Oconeelufty and Johnson's creek, the very paradise of trout fishers. Ready means may be had here also for longer trips, if desired ; as, for instance, through Heywood and Jackson counties to Franklin, in Macon county, the Nantahala valley, and the country of the Cherokees, a band of whom still remain in their old haunts, though the mass of their brethren is beyond the Mississippi. The remote parts of the mountains . . . abound in game—pheasants, turkeys, deer, wild-cats, even bears and wolves. A party of four, with a tent, a pack-horse, and a servant, at a moderate expense, and with great comfort and satisfaction, might spend a week or two in such excursions, enjoy fine sport, see scenery of exceeding interest and beauty, enjoy the delicious air of these mountains, and gain the health and energy lost in the toilsome pursuits of every-day occupations and harassments. The mountaineers are kind-hearted and hospitable, and the country, save in remote places, is sufficiently settled up to afford all necessary supplies which the gun and rod could not furnish."

Of main interest to my readers, however, will be the information that the long celebrated Warm Springs hotel, which stood in a beautiful meadow-glade, some fifty miles down the French Broad from Asheville, at the southern base of the Great Smoky, or Unaka mountains, and which was burned during the winter of 1884-5, is about to be rebuilt in such a way that it can accommodate better than ever the hosts of visitors who find its surroundings as near an elysium as can reasonably be expected in this world. Prue agrees with me, and adds that I must be sure to tell you to re-read " The Land of the Sky," before you go thither.

While there is everything to interest in the far away landscapes unrolled before the traveler's eye between Morristown and Knoxville, he sees that the district through which the railway runs is less populous and of poorer quality, as farming or timber land, than that between Morristown and Bristol. The general grade is a descending one, Bristol being 1,678 feet above the sea; Greenville, 1,581; Morristown 1,283; and Knoxville 900—a decrease which continues in about the same ratio to Chattanooga. The quality of the land is recovered when we reach the valley of the Holston, a broad dignified stream, pouring steadily on to join the French Broad, a few miles below, and form the Tennessee. Zinc, iron and a beautiful pink-veined marble occur in the ridges of this district, where there are broad corn fields in the bottoms and rich pastures "upon a thousand hills." Along this water course, which forms a straight avenue up and down the great valley, ran the "Cherokee trail" or Indian highroad; and almost where the track now passes, journeys of peace and trading, and expeditions of war and rapine were conducted by the red men, who have left only their musical names and crumbling earthworks as monuments of their former dominion.

Knoxville is the chief town of East Tennessee and one of growing wealth and importance. From a village of five or six thousand people at the end of the civil war, she has grown to a city of nearly twenty-five thousand, and a head-quarters of unusual enterprise. Here are the superintending offices of the great railway system we have been following, whose ramifications extend to Florida, Louisiana, and the western end of Tennessee. Here, too, comes in the Ohio division formed by the roads from Louisville and Cincinnati, which converge at Jellico, on the Kentucky state-line, and constitute what is called the "Jellico Route" between the South and the North-west. This road crosses the jumbled ridges and remote, sequestered valleys of the Cumberland mountains, through the gaps made by the Clinch river, Cove creek, and the charming Elk valley. It is a wild, almost unknown region, and one of those by-ways which we like now and then to seek out and enjoy alone. Southward from Knoxville a road, projected to penetrate eastern Georgia, has progressed some sixteen miles toward Chilhowee mountain. Here, also, is practically the head of navigation on the river for steamboats, to load which, flatboats and rafts bring cargoes from forests, farms and quarries far above.

These varied means of transportation have caused to be placed at Knoxville extensive iron works, car factories and machine shops of various kinds, which employ a large number of skilled mechanics. The population of the city is about half of northern or of foreign birth, and an air of activity and modern progressiveness animates the whole community. Woolen mills are already in operation, and a large cotton mill and woolen mill combined is under erection. Dozens of minor enterprises of the same character might be enumerated. As a trading town Knoxville has a peculiar prominence considering its size and situation. One firm alone of general merchants is said to do a business amounting to nearly two millions of dollars annually. A dozen houses exist in the city whose trade is wholly that of supplying country merchants with goods at wholesale, and they are enabled to stand between the retailer and eastern houses. Retail shops are numerous and well-filled, and agencies of every kind flourish. Gay street, the main business thoroughfare, is one of the most evenly and imposingly built-up avenues I know of in the country; a street of which the city has just right to be proud as typifying the completeness and solidity of its growth. To the man interested in the material progress of this region, Knoxville is most instructive.

The city is an extremely pleasant one, too. Its climate is charming, especially in winter, and its soil and situation most favorable for producing fine effects in architecture and gardening. On bold bluffs that steeply overlook the Tennessee river—a stream of inspiring stateliness and beauty, bringing with it the message of a hundred fountain-pregnant hills, and carrying the imagination fondly onward to its eternity in the space of the undying sea—are set, not too densely, the homes of the wealthy people of the town : homes replete with comfort and high culture. On a higher hill, remote from the thick of the town, and commanding a long extent of river and river cliffs, stands the University of Tennessee, which has a most elevating influence on the city. Just back of it is another hill upon which the town is slowly encroaching, where stood the heavy earthworks of Fort Saunders. Here, in 1864, occurred a bloody battle, in the storming of the hill and the fort by Longstreet's army, which found the defences too strong, and were compelled to retreat with great loss. Knoxville was for a long time the head-quarters of the Federal army in East Tennessee, and many of its citizens (as also the great proportion of the mountain people) remained loyal to the Union from first to last.

Looking northward from Fort Saunders, a large area of crowded town is seen lying out in the sunshine on the plain and knolls a mile or so back from the river. This is the newer part of the city, where it has outgrown the old limits and has not had time to spread the shade trees and cultivate the graces of the older portion. Many fine streets may be found over there, however, and a western briskness characterizes the appearance of things.

Seen from some high point like the garden of Mr. Dickinson's " Island " farm (which is a marvel of scientific agriculture, horticulture and stock-raising in the South, and should be neglected by no visitor), the country about Knoxville is, perhaps, as beautiful as can anywhere be found in the United States. A broad river with rocky bluffs and treegrown margin; highly cultivated areas, dotted with copses and isolated trees of countless species; patches of dark woodland, rising and falling with the irregular undulations of the rugged surface; and lastly the lofty and sublimely proportioned mountains which form so glorious a background;—all these surely might be pleasant, yet not fulfill the claim I have made. But here they are so gracefully disposed and related to one another, so rich in color, and " broad " in their arrangement and effect, that if one feature were omitted all the rest would be greatly impaired. But, to crown all, there is constantly in the atmosphere a moisture or some sort of softening quality which etherealizes, rounds off and makes gentle, soft and delicate, every object of the landscape it touches, until you never weary of Nature's face or cease to be soothed and fascinated by her loveliness.

Westward of Knoxville the country gradually becomes more open and level, though the many heights of the complicated Unaka or Great Smoky range still tower blue and very mountain-like in the south-east, while the north-western sky rests upon the wooded ramparts of Walden's ridge—an extension of the Cumberland mountain, which reaches diagonally across the whole state. At Loudon, twenty-eight miles below Knoxville, the "Tánisee " is crossed upon an iron bridge eighteen hundred feet in length, giving a lovely view from its height. Henceforth the river is always west of the track. A short distance above the bridge the Little Tennessee enters from the south. A small steamer plies from Loudon to Kingston, a farming town at the mouth of the Clinch. Great quantities of grain, brought by river steamers, are shipped upon the cars at Loudon and Kingston. Sweetwater, Mouse creek, Athens, and Riceville are stations of similar character, deriving their business from the farming region surrounding them. Athens has a newspaper, which has made a wide reputation, and its editor boasts of his town in this seductive style:—

She has the most genial climate of the earth; the most substantial court house in the state, the Wesleyan University, Athens Female Seminary, seven churches, cotton mills, woolen and flouring mills, and the prettiest girls under heaven's blue dome. In short, we are a God-blessed set, worshiping under our own vine and fig-tree, hanging the latch string on the outside, and inviting the world to come and enjoy with us our happiness.

Sixteen miles from Athens, over a fine mountain road, take one to the White Cliff Springs, three thousand feet above the sea and, therefore, in a pure and invigorating atmosphere, and enjoying a wide outlook. Between Calhoun and Charleston, the beautiful Hiwassee is crossed, and at Cleveland we halt for a dinner resembling Wordsworth's sweetheart:

"None knew thee but to love thee,
None named thee but to praise."

Cleveland is a thriving town of some two thousand inhabitants and contains much wealth, as is attested by the unusual elegance of its public buildings and mansions and the well-regulated appearance of its streets and sidewalks—a matter too often neglected in southern villages. Fine roads radiate from it through a lovely country, and the accommodations for visitors are good. Cleveland, consequently, is coming to be a favorite town for summer visitors from the far South, and winter residents escaping the chill of the North. The Ducktown copper mines are about forty miles distant; the road to them passes for twenty miles along the picturesque gorges of the Ocoee river, where it struggles out of the great mountains.

Cleveland is the point where the two great arms of the East Tennessee Virginia and Georgia Railway diverge. One arm reaches westward via Chattanooga, Decatur, and Corinth to Memphis and Trans-Mississippi connections. The other reaches southward to Rome, Georgia, and there divides: one branch proceeding through Atlanta and Macon to Brunswick, Savannah and Jacksonville; the other crossing Alabama via Calera and Selma to Meridian, and so on to Texas, and also via Calera and Montgomery to Mobile and New Orleans.

After a few words about the city of Chattanooga, which we have reached over the roughly wooded and rocky knolls about Oeltewah, and by tunneling underneath the blood-stained bulwark of Mission ridge, a final chapter of this little book will be devoted to explaining this southern system of transportation routes.

Before the war of the Secession Chattanooga was a miserable, muddy little village with an iron forge or two and some river-trading, but of no account to itself or the world in general. Seized upon as a strategic point and depot of supplies by the Confederates, when that war first broke out, it was fortified with the greatest possible strength and thought impregnable. Apprised of its danger by the approach of the army of the Cumberland over the ruins of strongholds in Kentucky and northern Tennessee, startled by the shock of Chickamauga and nerved to a last stand in Tennessee, it witnessed the terrific series of battles, of which those at Mission ridge and Lookout mountain were the most prominent episodes, and finally saw its defenders swept back from its forts.

Transformed in a day from a disloyal to a loyal stronghold, it formed scarcely more than a camp and quarter-master's depot for many months. Finally the flood of soldiers retreated and left Chattanooga more miserable and muddy than ever. But among those conquerors were shrewd men. When the war was over they went back there, bought town lots, farms, and mining rights. The citizens returned and help came from friends at other points. The qualities of situation and surrounding which had made it long ago a centre of Indian trails and traffic, which had caused armies to contest for its possession as a basis of military operations, now presented it to the business man as the most favorable spot in the district for his commercial and manufacturing schemes. Railways converged there; agents of big concerns north and south made it general head-quarters ; wholesale merchants competed for the local trade of a wide area; and the close proximity and easy "haul " of coal and minerals caused iron furnaces and rolling mills, metal factories and shops of every sort to be set up. Three-fourths at least of the citizens now (the population is nearly 25,000) are northern ; the town is new and growing with rapid strides ; the old-fashioned incommodious structures of the ante-bellum village have disappeared, and a rough, vigorous, active city has arisen, which is more like a town in Colorado than in sleepy South Tennessee. Thus far, as I say, the roughness of new construction and the vigor of business-haste is predominant in Chattanooga, and she is by no means lovely ; but many fine buildings have been or are being erected ; streets of ornate residences crown the heights above the noisy plain; and in a few years, under the fostering care of wealth and intelligence, and in the gracious climate which belongs to that region, Chattanooga will have grown out of this awkwardness into assured strength and admirable appearance.

XVIII.
A CHAPTER EXPLANATORY.

Memphis and Charleston Railway.—Blue Mountain Route to New Orleans and Mexico.—Florida Short Line.—Farewell and Bon Voyage.

It is the peculiarity of the South, as a whole, that its large towns are mainly placed along the coast or else under the edge of the mountains. Travelers whose destination is some small interior point have rarely a choice of routes, but must take that which carries them nearest. The main body of travelers, however, go from the North to some one of the southern seacoast cities or else come out of the South bound to one or the other of the great centres of civilization in the North. There is a third, and growing, class of travelers, who pass through the South *en route* to the far West, or in returning from Texas, Mexico and the Pacific, who wisely prefer the ever-varied scenery and pleasant climate of the Southern mountain region, and its opportunity of stopping at Washington and famous resorts, to the tiresome monotony of the Prairie states. To these "through" travelers, a few words of explanation in respect to the main routes traversing the South will well round out this little book

We have been facing southward and will continue so, standing at Chattanooga, on the very border of the Cotton and Gulf states. Supposing we were going to Memphis, and, perhaps, westward, there is open to us a route over the tracks of the Memphis and Charleston Railway which follows the Tennessee until it turns northward at the corner of Mississippi, and then passes on to Memphis through Corinth and Moscow. At Memphis a road may be taken straight across to Little Rock, Ark., where it meets the Missouri Pacific Railway system (via Hot Springs, Ark.), to middle and northern Texas, forming the most direct route between New York, New England and Texas or Mexico. There is also a very direct and pleasant railway track from Memphis to Kansas City, and to the prairies of southern Kansas beyond which the traveler may keep straight on by that most interesting of all "trails" to the Pacific coast, the "Santa Fé Route," which goes by the way of Pueblo and Salt Lake City through the heart of the Rocky mountains.

If, however, Mexico, southern Texas or Louisiana be the destination or starting point, then a choice of two paths is offered, converging at Calera, near the centre of Alabama, and labeled the "Blue Mountain Route" in the advertisements of the ticket agents. Leaving Chattanooga (or Cleveland), Tenn., the road at first passes through the hilly country of northern Georgia, where the names Cohutta, Dalton, Plainville, Kenesaw mountain and Rome will recall the sturdy fighting between the inevitable Sherman and the stubborn Hood. Passing out of these rugged hills into the more level but still pleasant country of north-eastern Alabama, the road skirts the iron district of that state (passing through the large new town of Anniston, where a connection is made for Birmingham), and pushes straight south-westward to Selma. This is a centre of much importance. From Selma railroads diverge to Montgomery, to Pensacola, to Mobile and eastward to Meridian, Miss. Meridian is another centre westward; a railway goes "straight as a string" to Jackson (whence a branch to Natchez), to Vicksburg, Shreveport and Marshall, connecting at the last point with the railway network of the Lone Star state. Some passengers may prefer this route to Texas over that by the way of Memphis. It is, perhaps, rather more interesting and there is no great difference, if any, in the time or expense. On both, the trains make fast time, run Sundays, are well equipped, provided with Pullman sleeping cars, and require no change of day coaches between Memphis and Chattanooga or Selma and Cleveland.

To go to New Orleans two courses are open. From Cleveland or Chattanooga one route is (as before) via Rome, Ga., and Anniston, to Calera, Ala.; thence you may go on to Meridian and down to New Orleans by the just completed New Orleans and North Eastern. Or, you may diverge at Calera and reach New Orleans by the old route through Montgomery (the capital of Alabama, and former seat of the Confederate government) Pensacola Junction (giving access to the famous naval station) and Mobile, the fashionable metropolis of the state. This way go Pullman sleeping cars and a most convenient arrangement of coaches. It takes in the largest towns, and, between Mobile and New Orleans, passes along the Gulf coast of Mississippi, giving charming glimpses of the sea-shore scenes and life, more like those of Italy than the Atlantic coast of the United States. By the way of Meridian, on the other hand, one passes through the very heart of the cotton and cane-brake region, and thus becomes acquainted with the sentiment and appearance of that remote, unruffled, agricultural life which formed the career of the greater part of the South under the old order of things.

All of these roads which pass through Meridian, Calera, and Rome to Cleveland, are united in a traffic arrangement of through tickets and through cars under the appropriate name of the " Blue Mountain Route," since for a thousand miles the traveler by it from Georgia northward, is within sight or actually upon either the Blue Ridge, or the azure Alleghanies. At one end is *Texas* and *New Orleans;* at the other *Cincinnati* (via Knoxville and Jellico), and *New York*, via the Shenandoah Valley.

There is another region in the South, however, to which Northern eyes turn when Winter's blustering winds beat upon a shrinking world. I mean Florida—called the Land of Flowers by Ponce de Leon, but which one might more truthfully name the Land of Winter Sunshine. Many of these travelers are invalids. To them speed and comfort are qualities of the highest importance in the chosen means of transportation. Suppose again that the traveler has reached Cleveland from Chicago, Cincinnati or Louisville by the way of Jellico; or from the North-east by the way of the Shenandoah; or that he has come to Chattanooga from the North-west. He is now at the fork of the roads—at the portal to the Florida his eyes foresee. The quickest way is to him the best. Georgia has not much to show in the way of scenery. When one gets south of the Kenesaw heights, where Sherman fought his terrible way to Atlanta, the landscape becomes level and the interest agricultural. Quickness and comfort, then, as I say, are the desiderata. As for rate of speed per mile, and the ease of a Pullman car, one first-class railway is much like another ; but the advantages of the line which goes by the way of Atlanta, Macon and Jesup, are that its course is more than 100 miles shorter than any other route (as one can easily see by a glance at the map), and that it is able to run both its Pullman sleeping cars (all the year round) and its passenger coaches from Cleveland and Chattanooga to Jacksonville without change—a ride of twenty-four hours. This, then, has justly assumed the name of the " Florida Short Line "—talismanic words at the ticket office !

Subsidiary advantages are, that this route permits a halt at Atlanta " the Chicago of the South," whence half a dozen local roads diverge. Then it follows for some time Sherman's " March to the Sea," always interesting. At Macon, east-and-west roads cross the state in four directions. At Jesup, Savannah and Brunswick may be reached by changing cars; while at Waycross, a road diverges across the southern margin of Georgia into western Florida and the cotton and orange region between Tallahassee and Mobile.

In review, then:

To Memphis, Little Rock, Hot Springs and northern Texas; or to

Kansas, Colorado, and the Pacific coast via Pueblo (or *vice versa*), the only path is over the Memphis and Charleston Railroad—that is, via *Chattanooga*, *Decatur* and *Corinth*.

To Texas, Mexico, and New Orleans, the advisable course is by one or another division (via Mobile or via Meridian) of the "Blue Mountain Route," that is, by the way of *Calera*.

To Florida, every consideration favors the "Florida Short Line," via *Atlanta* and *Jesup*.

And so—Farewell, and a pleasant journey to you!

WINTER EXCURSIONS

BY THE

Shenandoah Valley & Kennesaw Routes,

TO JACKSONVILLE, FLA.

No. 1.

FROM NORFOLK, VA., PETERSBURG, VA., BURKEVILLE, VA., LYNCHBURG, VA., and ROANOKE, VA.

No. 2.

FROM HAGERSTOWN, MD., SHENANDOAH JUNCTION, WEST VA., LURAY, VA., RIVERTON, VA., PORT REPUBLIC, VA., and WAYNESBORO JUNCTION, VA.

Route 1.—Norfolk & Western R. R. to Bristol; East Tennessee, Virginia & Georgia R. R. to Jesup; Savannah, Florida & Western R. R. to Jacksonville.

Route 2.— Shenandoah Valley R. R. to Roanoke; Norfolk & Western R. R. to Bristol; East Tennessee, Virginia & Georgia R. R. to Jesup; Savannah, Florida & Western R. R. to Jacksonville.

From RICHMOND.

Richmond & Alleghany R. R. to Lynchburg, thence to destination as above. ROUTE No. 1.

or

Richmond & Petersburg R.R. to Petersburg, thence to destination as above. ROUTE No. 1.

From NEW YORK, PHILADELPHIA & PENNSYLVANIA R. R. POINTS, Via HARRISBURG.

Pennsylvania R. R. to Harrisburg; Cumberland Valley R.R. to Hagerstown; thence to destination as above. ROUTE No. 2.

From BALTIMORE.

Western Maryland R. R. to Hagerstown; thence to destination as above. ROUTE No. 2.

or

Baltimore & Ohio R. R. to Shenandoah Junction, thence to destination as above. ROUTE No. 2.

From WASHINGTON and BALTIMORE & OHIO R. R. POINTS.

Baltimore & Ohio R. R. to Shenandoah Junction, thence to destination as above. ROUTE No. 2.

From NEW YORK, PHILADELPHIA, WILMINGTON and BALTIMORE, via WASHINGTON.

Pennsylvania R. R. to Washington; Virginia Midland to Lynchburg, thence to destination as above. ROUTE No. 1.

From WASHINGTON, CHARLOTTESVILLE, and VIRGINIA MIDLAND STATIONS.

Virginia Midland R. R. to Lynchburg, thence to destination as above. ROUTE No. 1.

LIST OF AGENTS

OF THE

SHENANDOAH VALLEY ROUTE AND KENNESAW ROUTE,

WHO WILL FURNISH TOURISTS GUIDE-BOOKS, TIME-TABLES, AND ALL INFORMATION OF RATES, ROUTES, TICKETS, SLEEPING-CAR RESERVATIONS, ETC., ETC.

C. P. GAITHER, Agt................290 Washington St., Boston, Mass.
H. V. TOMPKINS, East. Pass. Agt...........303 Broadway, New York.
B. H. FELTWELL, Pass. Agt.......838 Chestnut St., Philadelphia, Pa.
C. M. FUTTERER, Pass. Agt........................Hagerstown, Md.
W. H. FITZGERALD, Agt ....157 W. Baltimore St., Baltimore, Md.
E. J. LOCKWOOD, Pass. Agt......507 Penna. Ave., Washington, D. C.
ALLEN HULL, Pass. Agt ....Roanoke, Va.
T. H. BRANSFORD, Agt...............................Roanoke, Va.
J. F. CECIL, Agt.....................................Norfolk, Va.
WARREN L. ROHR, Ticket Agt............ ....Lynchburg, Va.
W. C. CARRINGTON, Ticket Agt....... ....Bristol, Tenn.
J. M. SUTTON, Pass. Agt.............. ....Chattanooga, Tenn.
JAMES MALOY, Pass. Agt.. .... Atlanta, Ga.
L. A. JETER, Ticket Agt...............................Macon, Ga.
B. H. HOPKINS, Pass. Agt...Cor. Bay & Hogan Sts., Jacksonville, Fla.
R. H. HUDSON, Pass. Agt.........................Montgomery, Ala.
J. C. ANDREWS, Gen. South'n Agt...158 Common St., New Orleans, La.
EUGENE SUTCLIFFE, Pass. Agt.....................Memphis, Tenn.
P. R. ROGERS, W. Pass. Agt......................Little Rock, Ark.

A. POPE, *Gen'l Pass. and Ticket Agent*,
NORFOLK & WESTERN and SHENANDOAH VALLEY R. Rs.,
ROANOKE, Virginia.

B. W. WRENN, *Gen'l Pass. and Ticket Agent*,
E. T. V. & G. R. R., KNOXVILLE, Tennessee.

DIRECTORY OF AGENCIES.

WHERE THROUGH TICKETS—BOTH STRAIGHT AND ROUND-TRIP- FLORIDA, NEW ORLEANS AND SUMMER EXCURSION—ARE SOLD, INFORMATION GIVEN, TIME-CARDS FURNISHED, AND SLEEPING-CAR BERTHS AND SECTIONS RESERVED TO ALL POINTS ON OR VIA THE RAILWAYS OF THE

SHENANDOAH VALLEY AND KENNESAW ROUTES.

IN THE NORTH AND EAST.

BOSTON, at No. 3 Old State House; 205, 211, 214, 232 and 322 Washington Street; and at the Depots of the New York Lines, and Office of Line, 290 Washington Street.

ALSO, at Railroad Ticket Offices at Providence, Worcester, Springfield, Hartford New Haven, Bridgeport, Stamford, etc.

NEW YORK, at No. 1 Astor House; No. 8 Battery Place; 315, 435, 849 and 943 Broadway; and 168 East 125th Street; Depots foot of Desbrosses and Cortlandt Streets, and Office of Line, 303 Broadway.

BROOKLYN, at No. 4 Court Street, and Office of Brooklyn Annex, foot of Fulton Street.

JERSEY CITY, at Penn. R. R. Depot Ticket Office; also, at Passenger Station Ticket Offices, Penn. R. R. at Newark, Elizabeth, Rahway, New Brunswick, and Trenton, N. J.

PHILADELPHIA, at Nos. 838, 1100 and 1348 Chestnut Street; and at Depot, Broad and Market Streets; also, at R.R. Ticket Offices Penn R. R., at Germantown, Pa., Chester, Pa., Wilmington, Del.

HARRISBURG, at Ticket Office, Cumberland Valley R. R.

PITTSBURG, at Depot Ticket Offices.

BALTIMORE, at Ticket Office, Western Maryland R. R., 133 West Baltimore Street; at Depot Western Maryland R. R.; and Office of Line, 157 West Baltimore Street.

WASHINGTON, at Depot of the Baltimore & Ohio R. R.; at Depot Penn. R. R.; also, 601 Penn. Ave.; and Office of the Line, 507 Pennsylvania Avenue.

NORFOLK, at Office, W. T. Walke, Ticket Agent, under Atlantic Hotel; W. I. Flournoy, Ticket Agent, Purcell House; also, at Depot N. & W R. R.

RICHMOND, at Depot Richmond & Petersburg, Richmond & Danville, and Richmond & Alleghany R. Rs.; also, at 1000 Main Street, A. W. Garber & Co., General Agents, 1200 Main Street. S. H. Bowman, Ticket Agent.

HAGERSTOWN, MD., at Ticket Office, Shenandoah Valley R. R.

ROANOKE, VA., at Depot Shenandoah Valley and Norfolk & Western Railr'ds.

And at Coupon Ticket Offices of all lines connecting at Harrisburg, Washington, Hagerstown and Shenandoah Junction.

IN THE SOUTH AND SOUTHWEST.

ATLANTA, GA., at Ticket Office, Depot East Tenn., Va. & Ga. R. R., and W. & A. R. R. Ticket Office.

CHATTANOOGA, TENN., at Depot Ticket Office E. T. V. & G. R. R.

MACON, GA., at Depot Ticket Office, and at 102 Mulberry Street.

JACKSONVILLE, FLA., at Ticket Office S. F. & W. R. R., and Office of Line, corner West Bay & Hogan Streets.

ST. AUGUSTINE, FLA., Ticket Office, S. F. & W. R. R.

SAVANNAH, GA., at Ticket Office S. F. & W. R. R., and Central R. R. of Ga.

VICKSBURG, MISS., at Depot Ticket Office, V. & M. R. R.

MERIDIAN, MISS. at Depot Ticket Offices E. T. V. & G. R. R., and Ala. Gt. So. R. R.

SELMA, ALA., at Depot Ticket Office E. T. V. & G. R. R.

MONTGOMERY, ALA., at Depot of West Ala. R., and L. & N. R. R.

MOBILE, ALA., at Ticket Office Battle House, and Depot Ticket Offices L.& N. R.R. and M. & O. R. R.

NEW ORLEANS, LA., at Ticket Offices and Depots of L. & N. R. R., Illinois Central R. R., N. O. & N. E. R. R., and Office of Line, 158 Common Street.

GALVESTON, TEXAS, at 116 Tremont Street, and Depot Ticket Offices G. H. & H. R. R.

HOUSTON, TEXAS, at Depot Ticket Offices T. & N. O. R. R., and I. & Gt. N. R. R.

SAN ANTONIO, TEXAS, at Ticket Office and Depot of G. H. & S. A. R. R.

MEMPHIS, TENN., at Main Street Ticket Office, Barney Hughes, Ticket Agent - and Depot M. & C. R. R.

LITTLE ROCK, ARK., at Depot Ticket Office M. & Little Rock R. R.

TEXARKANA, TEXAS, at Depot Ticket Office St. L. & I. Mt. R. R.

DALLAS, TEXAS, at Depot Ticket Office Texas & Pacific R. R.

AND AT TICKET OFFICES OF ALL CONNECTING LINES.

ITINERARY OF ROUTES

By which Florida, New Orleans, Luray, Natural Bridge, and the Noted Summer Excursion Resorts of Virginia are Reached Quickest, Cheapest and Best.

BY

The Shenandoah Valley Route.—The Kennesaw Route,

AND THEIR CONNECTING RAILWAY LINES,

ALSO,

To New Orleans and the World's Exposition!!

GROUP A.

From NORFOLK, PETERSBURG, BURKEVILLE, FARMVILLE, LYNCHBURG, LIBERTY, ROANOKE, MARION, ABINGDON, and other Principal Towns and Cities.

GROUP B.

From HAGERSTOWN, SHEPHERDSTOWN, SHENANDOAH JUNCTION, BOYCE, LURAY, RIVERTON, WAYNESBORO JUNCTION, NATURAL BRIDGE, BUCHANAN, Etc., Etc., Etc.

ROUTES—GROUP A.

No. 1.—Norfolk & Western R. R. to Bristol; East Tenn., Va. & Ga. R. R. to Calera; Louisville & Nashville to New Orleans.

No. 2.—Norfolk, & Western R. R. to Bristol; East Tenn., Va. & Ga. R. R. to Atlanta; A. & W. Point R. R. to West Point; Western Railway of Ala. to Montgomery; Louisville & Nashville R. R. to New Orleans.

No. 3.—Norfolk & Western R. R. to Bristol; East Tenn., Va. & Ga. R. R. to Dalton; Western & Atlantic R. R. to Atlanta; A. & West Point R. R. to West Point; Western Railway of Ala. to Montgomery; L. & N. R. R. to New Orleans.

No. 4.—Norfolk & Western R. R. to Bristol; East Tenn., Va. & Ga. R. R. to Chattanooga; Queen & Crescent Route to New Orleans.

No. 5.—Norfolk & Western R. R. to Bristol; East Tenn., Va. & Ga. R. R. to Chattanooga; Memphis & Charleston to Grand Junction; Illinois Central to New Orleans.

ROUTES—GROUP B.

No. 1.—Shenandoah Valley R. R. to Roanoke; Norfolk & Western R. R. to Bristol; East Tenn., Va. & Ga. R. R. to Calera; Louisville & Nashville R. R. to New Orleans.

No. 2.—Shenandoah Valley R. R. to Roanoke; Norfolk & Western R. R. to Bristol; East Tenn., Va. & Ga. R. R. to Atlanta; Atlanta & West Point R. R. to West Point; Western Railway of Ala. to Montgomery; Louisville & Nashville R.R. to New Orleans.

No. 3.—Shenandoah Valley R. R. to Roanoke; Norfolk & Western R. R. to Bristol; East Tenn., Va. & Ga. R. R. to Dalton; Western & Atlantic R. R. to Atlanta; Atlanta & West Point R. R. to West Point; Western Railway of Ala. to Montgomery; Louisville & Nashville R. R. to New Orleans.

No. 4—Shenandoah Valley R. R. to Roanoke; Norfolk & Western R. R. to Bristol; East Tenn., Va. & Ga. R. R. to Chattanooga; Queen & Crescent Route to New Orleans.

No. 5.—Shenandoah Valley R. R. to Roanoke; Norfolk & Western R. R. to Bristol; East Tenn., Va. & Ga. R. R. to Chattanooga; Memphis & Charleston R. R. to Grand Junction; Illinois Central R. R. to New Orleans.

From RICHMOND.

Richmond & Alleghany R. R. to Lynchburg, thence to destination as per routes named in Group A.

or

Richmond & Petersburg R. R. to Petersburg, thence to destination as per routes given in Group A.

From NEW YORK, PHILADELPHIA and PENNSYLVANIA R. R. POINTS, Via HARRISBURG.

Pennsylvania R. R. to Harrisburg; Cumberland Valley R. R. to Hagerstown, thence to destination as per routes given in Group B.

From BALTIMORE.

Western Maryland R. R. to Hagerstown, thence to destination as per routes given in Group B.

or

Baltimore & Ohio R. R. to Shenandoah Junction, thence to destination as per routes given in Group B.

From WASHINGTON and BALTIMORE & OHIO R. R. POINTS.

Baltimore & Ohio R. R. to Shenandoah Junction, thence to destination as per routes given in Group B.

From NEW YORK, PHILADELPHIA, WILMINGTON AND BALTIMORE, via WASHINGTON.

Pennsylvania R. R. to Washington, Virginia Midland R. R. to Lynchburg, thence to destination as per routes given in Group A.

From WASHINGTON, CHARLOTTESVILLE and STATIONS ON VA. MIDLAND R. R.

Virginia Midland R. R. to Lynchburg, thence to destination as per routes given in Group A.

GROUP A.
Afton, Va.
Clifton Forge, Va.
Covington, Va.
Goshen, Va.
Greenbrier White Sulphur, W. Va.
Kanawha Falls, W. Va.
Millboro, Va.
Staunton, Va.

The above resorts are located immediately on line of Chesapeake & Ohio Railway, and are reached without staging.

GROUP B.
Bath Alum, Va.—*Millboro.*
Cold Sulphur, Va.—*Goshen.*
Hot or Healing, Va.—*Covington.*
Millboro Springs, Va.—*Millboro.*
Mountain Top House, Va.—*Afton.*
Rockbridge Baths, Va.—*Goshen.*
Salt Sulphur, W. Va.—*Fort Springs.*
Stribling, Va.—*Staunton.*
Sweet Chalybeate, Va.—*Alleghany.*
Walawhatoola, Va.—*Millboro.*

The above resorts are located *off* the line of Chesapeake & Ohio Railway. Station in *italic* type indicates point of departure from railroad and where stage must be taken.

GROUP C.
Dagger's Springs, Va.—*Gala Water.*
Rockbridge Alum Springs, Va.—*Lexington.*

Located *off* line of Richmond & Alleghany Railroad. Station in *italic* type indicates point of departure from railroad, and where stage must be taken.

GROUP D.
Abingdon, Va.
Big Springs, Va.
Big Tunnel, Va.
Blue Ridge Springs, Va.
Buford's,   "
Christiansburg   "
Dublin,   "
Egglestons,   "
Gishs,   "
Glade Springs,   "
Liberty,   "
Marion,   "
Montgomery White,   "
Roanoke,   "
Rural Retreat,   "
Salem,   "
Saltville,   "
Wytheville,   "

The above resorts are located on line of Norfolk & Western Railroad, and are reached without staging.

GROUP E.
Alleghany Springs, Va.—*Shawsville.*
Bedford Alum Springs, Va.—*Forest.*
Blacksburg Springs, Va.—*Christiansburg.*
Botetourt Springs, Va.—*Salem.*
Chillhowee Springs, Va.—*Greevers.*
Coyner's Springs, Va.—*Bonsacks.*
Farmville Lithia, Va.—*Farmville.*
Hunter's Alum Springs, Va.—*Dublin.*
Lake Springs, Va.—*Salem.*
Monroe Red Sulphur Springs, W. Va.—*Glen Lyn.*
Mountain Lake, Va.—*Staytide.*
Pulaski Alum Springs, Va.—*Dublin.*
Roanoke Red Sulphur Springs, Va.—*Salem.*
Seven Springs, Va.—*Glade Springs.*
Sharon Springs, Va.—*Wytheville.*
Washington Springs, Va.—*Glade Springs.*
Yellow Sulphur Springs, Va.—*Christiansburg.*

The above springs are located *off* the line of Norfolk & Western railroad. Station in *italic* type indicates point of departure from railroad, and where stage or hack must be taken.

GROUP F.
Berryville, Va.
Buchanan, Va.
Charlestown, W. Va.
Hagerstown, Md.
Luray, Va.
Roanoke, Va.
Shepherdstown, W. Va.
White Post, Va.

The above resorts are located on line of Shenandoah Valley railroad, and are reached without staging.

GROUP G.
Almirida, Va.—*Berryville.*
Baker's Springs, Va.—*Waynesboro.*
Botetourt, Va.—*Cloverdale.*
Bon Air, Va.—*Elkton.*
Fincastle Mineral, Va.—*Cloverdale.*
Natural Bridge, Va.—*Natural Bridge.*
Rockingham Springs, Va.—*Elkton.*
The Vineyard, Va.—*Boyceville.*

The above resorts are located *off* line of Shenandoah Valley railroad. Station in *italic* type indicates point of departure from railroad, and where stage or hack must be taken.

GROUP H.
Capon Springs, W. Va.—*Capon.*
Rawley Springs, Va.—*Harrisonburg.*
Shenandoah Alum Springs, Va.—*Mt Jackson.*
Orkney Springs, Va.—*Mt. Jackson.*

The above resorts are located *off* line of Valley Branch, Baltimore & Ohio railroad. The station in *italic* type indicates point of departure from railroad, and where stage or hack must be taken.

GROUP I.
Old Point Comfort, Va.

GROUP J.
Chambersburg, Pa.
Greencastle, Pa.
Mechanicsburg, Pa.
Shippensburg, Pa,

The above resorts are located on line of Cumberland Valley Railroad, and are reached without staging.

GROUP K.
Warm Springs, N. C.
Asheville, N. C.

The above resorts are located on line of Western North Carolina R. R., and are reached without staging.

SUMMER EXCURSION ROUTES,

From NORFOLK, VA.

To resorts named in GROUP A.
Norfolk & Western Railroad to Petersburg. Richmond & Petersburg Railroad to Richmond. Transfer to Chesapeake & Ohio Depot. Chesapeake & Ohio Railway to destination.

To resorts named in GROUP B.
Norfolk & Western Railroad to Petersburg. Richmond & Petersburg Railroad to Richmond. Transfer to Chesapeake & Ohio Depot. Chesapeake & Ohio Railway to nearest station. Stage to destination.

To resorts named in GROUP C.
Norfolk & Western Railroad to Lynchburg. Richmond & Alleghany to nearest station. Stage to destination.

To resorts named in GROUP D.
Norfolk & Western Railroad to destination.

To resorts named in GROUP E.
Norfolk & Western Railroad to nearest station. Stage or hack to destination.

To resorts named in GROUP F.
Norfolk & Western Railroad to Roanoke. Shenandoah Valley Railroad to destination.

To resorts named in GROUP G.
Norfolk & Western Railroad to Roanoke. Shenandoah Valley Railroad to nearest station. Stage to destination.

To resorts named in GROUP H.
Norfolk & Western Railroad to Petersburg. Richmond & Petersburg Railroad to Richmond. Transfer to Chesapeake & Ohio Depot. Chesapeake & Ohio Railroad to Staunton. Valley Branch, Baltimore & Ohio to nearest station. Stage to destination.

or

Norfolk & Western Railroad to Roanoke. Shenandoah Valley Railroad to Waynesboro'. Chesapeake & Ohio Railway to Staunton. Valley Branch, Baltimore & Ohio Railroad to nearest station. Stage to destination.

To resorts named in GROUP J.
Norfolk & Western Railroad to Roanake. Shenandoah Valley Railroad to Hagerstown. Cumberland Valley Railroad to destination.

To resorts named in GROUP K.
Norfolk & Western Railroad to Bristol. East Tennessee, Virginia & Georgia Railroad to Unaka. Western North Carolina Railroad to nearest station.

From PETERSBURG.

To resorts named in GROUP C.
Norfolk & Western Railroad to Lynchburg. Richmond & Alleghany Railroad to nearest station. Stage to destination.

To resorts named in GROUP D.
Norfolk & Western Railroad to destination.

To resorts named in GROUP E.
Norfolk & Western Railroad to nearest station. Stage or hack to destination.

To resorts named in GROUP F.
Norfolk & Western Railroad to Roanoke. Shenandoah Valley Railroad to destination.

To resorts named in GROUP G.
Norfolk & Western Railroad to Roanoke. Shenandoah Valley Railroad to nearest station. Stage or hack to destination.

To resorts named in GROUP H.
Norfolk & Western Railroad to Roanoke. Shenandoah Valley Railroad to Waynesboro. Chesapeake & Ohio Railroad to Staunton. Valley Branch, Baltimore & Ohio to nearest station. Stage or hack to destination.

OLD POINT COMFORT, VA.
Norfolk & Western Railroad to Norfolk. Bay Line steamer to Old Point.

To resorts named in GROUP K.
Norfolk & Western Railroad to Bristol. East Tennessee, Virginia & Georgia Railroad to Unaka. Western North Carolina Railroad to destination.

From WELDON, GOLDSBORO, RALEIGH (*via* Weldon), WILMINGTON, CHARLESTON, SAVANNAH, COLUMBIA (*via* W. C. & A. R.R.), JACKSONVILLE (*via* Charleston).

Take the Atlantic Coast Line to Petersburg, Va., thence to destination. (See routes from Petersburg.)

From RICHMOND, VA.

To resorts named in GROUP D.
Richmond & Petersburg Railroad to Petersburg. Norfolk & Western Railroad to destination.

or

Richmond & Alleghany Railroad to Lynchburg. Norfolk & Western Railroad to destination.

To resorts named in GROUP E.
Richmond & Petersburg Railroad to Petersburg. Norfolk & Western Railroad to nearest station. Stage or hack to destination.

or

Richmond & Alleghany Railroad to Lynchburg. Thence as above.

To resorts named in GROUP F.
Chesapeake & Ohio Railroad to Waynesboro. Shenandoah Valley Railroad to destination.

To resorts named in GROUP G.
Chesapeake & Ohio Railroad to Waynesboro. Shenandoah Valley Railroad to nearest station. Stage or hack to destination.

To OLD POINT COMFORT, Va.
Richmond & Petersburg Railroad to Petersburg. Norfolk & Western Railroad to Norfolk. Bay Line Steamer to Old Point.

To resorts named in GROUP K.
Richmomd & Petersburg Railroad to Petersburg. Norfolk & Western Railroad to Bristol. East Tennessee, Virginia & Georgia Railroad to Unaka. Western North Carolina Railroad to destination.
or
Richmond & Alleghany Railroad to Lynchburg. Thence as above.

To resorts named in GROUP J.
Chesapeake & Ohio Railroad to Waynesboro. Shenandoah Valley Railroad to Hagerstown. Cumberland Valley Railroad to destination.

From HAGERSTOWN, SHENANDOAH JUNCTION and RIVERTON JUNCTION.

To resorts named in GROUP A.
Shenandoah Valley Railroad to Waynesboro. Chesapeake & Ohio Railroad to destination.

To resorts named in GROUP B.
Shenandoah Valley Railroad to Waynesboro. Chesapeake & Ohio Railroad to nearest station. Stage or hack to destination.

To resorts named in GROUP D.
Shenandoah Valley Railroad to Roanoke. Norfolk & Western Railroad to destination.

To resorts named in GROUP E.
Shenandoah Valley Railroad to Roanoke. Norfolk & Western Railroad to nearest station. Stage or hack to destination.

To resorts named in GROUP F.
Shenandoah Valley Railroad to destination.

To resorts named in GROUP G.
Shenandoah Valley Railroad to nearest station. Stage or hack to destination.

To resorts named in GROUP H.
Shenandoah Valley Railroad to Waynesboro. Chesapeake & Ohio Railroad to Staunton. Valley Branch, Baltimore & Ohio Railroad to nearest station. Stage or hack to destination.

To OLD POINT COMFORT, Va.
Shenandoah Valley Railroad to Roanoke. Norfolk & Western Railroad to Norfolk. Bay Line steamer to Old Point.

To resorts named in GROUP K.
Shenandoah Valley Railroad to Roanoke. Norfolk & Western Railroad to Bristol. East Tennessee, Virginia & Georgia Railroad to Unaka. Western North Carolina Railroad to nearest station.

From NEW YORK, PHILADELPHIA and P. R. R. points, via Harrisburg.
Pennsylvania Railroad to Harrisburg, Pa. Cumberland Valley Railroad to Hagerstown. Thence to destination. (See routes from *Hagerstown, Md.*)

From BALTIMORE.
Western Maryland Railroad to Hagerstown. Thence to destination. (See routes from *Hagerstown, Md.*)

From WASHINGTON and Balt. & Ohio R. R. points, via B. & O.
Baltimore & Ohio Railroad to Shenandoah Junction. Thence to destination. (See routes from *Shenandoah Junction.*)

From WASHINGTON, via Va. Mid. R. R. & Manassas Branch.
Virginia Midland Railroad (Manassas Branch) to Riverton. Thence to destination. (See routes from *Riverton Junction.*)

From WASHINGTON and CHARLOTTESVILLE, via Lynchburg.
To resorts named in GROUP D.
Virginia Midland Railroad to Lynchburg. Norfolk & Western Railroad to destination.

To resorts named in GROUP E.
Virginia Midland Railroad to Lynchburg. Norfolk & Western Railroad to nearest station. Stage or hack to destination.

To resorts named in GROUP K.
Virginia Midland Railroad to Lynchburg. Norfolk & Western Railroad to Bristol. East Tennessee, Virginia & Georgia Railroad to Unaka. Western North Carolina Railroad to destination.

From NEW YORK, PHILADELPHIA and BALTIMORE, via Washington.
Pennsylvania Railroad to Washington. Thence to destination. (See routes from Washington, *via* Virginia Midland Railroad and Lynchburg.)

From DANVILLE, Va., GREENSBORO, RALEIGH, via Greensboro, SALISBURY, CHARLOTTE, Etc., via DANVILLE.

To resorts named in GROUP D.
Richmond & Danville Railroad to Danville. Virginia Midland Railroad to Lynchburg. Norfolk & Western Railroad to destination.

To resorts named in GROUP E.

Richmond & Danville Railroad to Danville. Virginia Midland Railroad to Lynchburg. Norfolk & Western Railroad to nearest station. Stage or hack to destination.

To resorts named in GROUP F & G.

Richmond & Danville Railroad to Danville. Virginia Midland Railroad to Lynchburg. Norfolk & Western Railroad to Roanoke. Shenandoah Valley Railroad to destination.

To OLD POINT COMFORT, Va.

Richmond & Danville Railroad to Burkeville. Norfolk & Western Railroad to Norfolk. Bay Line Steamer to Old Point.

From CHATTANOOGA, DALTON, CALERA, SELMA, CLEVELAND and KNOXVILLE, and points on line of E. T. V. & G. R.R.

To resorts named in GROUP A.

East Tennessee, Virginia & Georgia Railroad to Bristol. Norfolk & Western Railroad to Roanoke. Shenandoah Valley to Waynesboro. Chesapeake & Ohio to destination.

or

East Tenessee, Virginia & Georgia Railroad to Bristol. Norfolk & Western Railroad to Lynchburg. Virginia Midland Railroad to Charlottesville. Chesapeake & Ohio to destination.

To resorts named in GROUP B.

East Tennessee, Virginia & Georgia Railroad to Bristol. Norfolk & Western Railroad to Roanoke. Shenandoah Valley Railroad to Waynesboro. Chesapeake & Ohio Railroad to nearest station. Stage to destination.

or

East Tennessee, Virginia & Georgia Railroad to Bristol. Norfolk & Western Railroad to Lynchburg. Virginia Midland to Charlottesville. Chesapeake & Ohio Railroad to nearest station. Stage or hack to destination.

To resorts named in GROUP D.

East Tennessee, Virginia & Georgia Railroad to Bristol. Norfolk & Western Railroad to destination.

To resorts named in GROUP E.

East Tennessee, Virginia & Georgia Railroad to Bristol. Norfolk & Western Railroad to nearest station. Stage or hack to destination.

To resorts named in GROUP F.

East Tennessee, Virginia & Georgia Railroad to Bristol. Norfolk & Western Railroad to Roanoke. Shenandoah Valley to destination.

To resorts named in GROUP G.

East Tennessee, Virginia & Georgia Railroad to Bristol. Norfolk & Western Railroad to Roanoke. Shenandoah Valley Railroad to nearest station. Stage or hack to destination.

To resorts named in GROUP H.

East Tennessee, Virginia & Georgia Railroad to Bristol. Norfolk & Western Railroad to Roanoke. Shenandoah Valley Railroad to Waynesboro. Chesapeake & Ohio Railroad to Staunton. Valley Branch Baltimore & Ohio Railroad to nearest station. Stage or hack to destination.

or

East Tennessee, Virginia & Georgia Railroad to Bristol. Norfolk & Western Railroad to Lynchburg. Virginia Midland Railroad to Charlottesville. Chesapeake & Ohio Railroad to Staunton. Valley Branch, Baltimore & Ohio Railroad to nearest station. Stage or hack to destination.

To OLD POINT COMFORT, Va.

East Tennessee, Virginia & Georgia Railroad to Bristol. Norfolk & Western Railroad to Norfolk. Bay Line Steamer to Old Point.

To resorts named in GROUP J.

East Tennessee, Virginia & Georgia Railroad to Bristol. Norfolk & Western Railroad to Roanoke. Shenandoah Valley Railroad to Hagerstown. Cumberland Valley Railroad to destination.

From NASHVILLE, Tenn., and line of N. C. & St. L. R. R.

Nashville, Chattanooga & St. Louis Railroad to Chattanooga. Thence to destination as indicated in route from Chattanooga.

From MERIDIAN, YORK, BIRMINGHAM and line of A. G. S. R. R.

Alabama Great Southern Railroad to Chattanooga. Thence to destination as indicated in route from Chattanooga.

From ATLANTA, Ga.

Western & Atlantic Railroad to Dalton. Thence to destination as indicated in Routes from Dalton.

or

East Tennessee, Virginia & Georgia Railroad to Cleveland. Thence to destination as indicated in routes from Cleveland.

From MACON, Ga., and South Western Georgia points.

Central Railroad of Georgia to Atlanta. Western & Atlantic Railroad to Dalton.

or

East Tennessee, Virginia & Georgia Railroad to Cleveland, Tenn. Thence to destination as indicated in routes from Dalton and Cleveland.

From MONTGOMERY & OPELIKA, via Atlanta.

Western Railroad of Alabama to West Point. Atlanta & West Point Railroad to Atlanta. Western & Atlantic Railroad to Dalton.

or

Western Railroad of Alabama to West Point. Atlanta & West Point Railroad to

Atlanta. East Tennessne, Virginia & Georgia Railroad to Cleveland. Thence to destination as indicated in routes from Dalton and Cleveland.

From NEW ORLEANS, MOBILE, PENSACOLA and L. & N. R. R. points, via Atlanta.

Louisville & Nashville Railroad to Montgomery. Western Railroad of Alabama to West Point. Atlanta & West Point Railroad to Atlanta. Western & Atlantic Railroad to Dalton.

or

Louisville & Nashville Railroad to Montgomery. Western Railroad of Alabama to West Point. Atlanta & West Point Railroad to Atlanta. East Tennessee, Virginia & Georgia Railroad to Cleveland. Thence to destination as indicated in routes from Dalton and Cleveland.

From NEW ORLEANS, MOBILE, PENSACOLA, MONTGOMERY, and L. & N. R. R. points, via Calera.

Louisville & Nashville Railroad to Calera, Ala. Thence to destination as indicated in routes from Calera, Ala.

From NEW ORLEANS, VICKSBURG, JACKSON, via Grand Junction.

Illinois Central Railroad to Grand Junction. Memphis & Charleston Railroad to Chattanooga. Thence to destination, as indicated in route from Chattanooga.

In addition to the direct routes of travel given in the foregoing ITINERARY, which are in all cases the same in each direction, the entirely new feature in Excursion Travel of

VARIABLE ROUTES,

by which tourists going from home by one line may return by another, has been arranged; this being by reason of the extensive mileage of the

Virginia, Tennessee & Georgia Air Line,

traversing large areas of diverse territory—an entirely practicable arrangement within its own control.

These Variable Route Tickets embrace all or portions only of the Scenic Attractions and Summer Resorts of the Line, according to the TASTE, TIME and MEANS of intending tourists, and are obtainable during the Excursion Season at the offices of the Line, or initial companies at interest, in the following cities:

BALTIMORE.—Western Maryland Railroad, Hillen Station, Fulton Station, Pennsylvania Avenue Station, at 133 West Baltimore Street. Geigan & Co., ticket agents.

Baltimore & Ohio Railroad, Camden Station, and corner Baltimore & Calvert Streets.

Baltimore Steam Packet Company, 157 West Baltimore Street. W. H. Fitzgerald, agent.

WASHINGTON.—507 Pennsylvania Ave., E. J. Lockwood, Passenger agent. Baltimore & Ohio Railroad, Depot ticket office.

Virginia Midland Railroad ticket office, 601 Pennsylvania Avenue. N. McDaniel, ticket agent.

HARRISBURG.—Ticket office of Cumberland Valley Railroad.

HAGERSTOWN.— Ticket office of Shenandoah Valley Railroad. Charles Feldman, ticket agent.

LURAY.—Ticket office of Shenandoah Valley Railroad. M. Spitler, ticket agent

WAYNESBORO.—Ticket office of Shenandoah Valley Railroad. B. L. Greider, ticket agent.

NORFOLK.—Under Atlantic Hotel. W. T. Walke, ticket agent. Purcell House, W. I. Flournoy, ticket agent.

At Depot Norfolk & Western Railroad. J. F. Cecil, agent.

NEW YORK.—At office of the Line, 303 Broadway. H. V. Tompkins, agent.

BOSTON —At office of the Line, 290 Washington Street. C. P. Gaither, agent.

LYNCHBURG.—Norfolk & Western Railroad Depot ticket office. W. L. Rohr, ticket agent.

PETERSBURG.—Norfolk & Western Railroad Depot ticket office. H. V. L. Bird, agent.

RICHMOND.—At 1000 Main Street. A. W. Garber & Co., ticket agents, 1206 Main Street. S. H. Bowman, agent, R. & A. R. R.

ROANOKE.—Norfolk & Western and Shenandoah Valley Railroad Depot ticket office. T. H. Bransford, agent.

KNOXVILLE.—East Tennessee, Virginia & Georgia Railroad Depot ticket office.

CHATTANOOGA.—East Tennessee, Virginia & Georgia Railroad Depot ticket office. J. H. Peebles, ticket agent.

MEMPHIS.—Memphis & Charleston Railroad Depot ticket office. Also Main Street ticket office. Barney Hughes, ticket agent.

ATLANTA.—East Tennessee, Virginia & Georgia Railroad Depot ticket office. Jack W. Johnson, ticket agent.

MACON.—East Tennessee, Virginia & Georgia Railroad Depot ticket office. R. T. Reynolds, ticket agent. Also at 102 Mulberry Street. Burr Brown, ticket agent.

JACKSONVILLE. — Savannah, Florida & Western Railroad ticket office, West Bay Street. And office of the Line, corner Bay and Hogan Streets. B. H. Hopkins, passenger agent.

SELMA.—East Tennessee, Virginia & Georgia Railroad at Depot ticket office. T. H. Lavender, ticket agent.

MERIDIAN.—East Tennessee, Virginia & Georgia Railroad at Depot ticket office. C. Berney, ticket agent.

www.ingramcontent.com/pod-product-compliance
Lightning Source LLC
Chambersburg PA
CBHW031347160426
43196CB00007B/753